AF361672

Advanced Methods for Seismic Performance Evaluation of Building Structures

Advanced Methods for Seismic Performance Evaluation of Building Structures

Editor

Sang Whan Han

MDPI • Basel • Beijing • Wuhan • Barcelona • Belgrade • Manchester • Tokyo • Cluj • Tianjin

Editor
Sang Whan Han
Hanyang University
Korea

Editorial Office
MDPI
St. Alban-Anlage 66
4052 Basel, Switzerland

This is a reprint of articles from the Special Issue published online in the open access journal *Applied Sciences* (ISSN 2076-3417) (available at: https://www.mdpi.com/journal/applsci/special_issues/seismic_performance_building_structures).

For citation purposes, cite each article independently as indicated on the article page online and as indicated below:

LastName, A.A.; LastName, B.B.; LastName, C.C. Article Title. *Journal Name* **Year**, *Article Number*, Page Range.

ISBN 978-3-03943-214-1 (Hbk)
ISBN 978-3-03943-215-8 (PDF)

Contents

About the Editor

Sang Whan Han has been working for Hanyang University, Seoul, Korea, as a professor since 1997. Prior to joining the faculty of Hanyang University, he studied at the University of Illinois at Urbana-Champaign. He received his master's and Ph.D. in 1990 and 1994, respectively. He earned his bachelor's degree in 1986 from Hanyang University, Seoul, Korea. He has been active in the earthquake engineering field since his studies in graduate school. Currently, he is a vice president of the Korean Society of Earthquake Engineering. He is also a licensed engineer in Korea and the US: a professional engineer in the US and a structural engineer in Korea. His research topics are (1) seismic design of tall buildings, (2) development of seismic performance evaluation procedures, (3) structural reliability analysis, and (4) nonlinear analysis of large structures. He has been involved in many academic projects focusing on earthquake engineering, as well as engineering consulting projects in the field of earthquake engineering.

Preface to "Advanced Methods for Seismic Performance Evaluation of Building Structures"

Earthquakes are one of the most dangerous natural events, inflicting damage and leading to the collapse of buildings and infrastructures. On average, 10,000 people lose their lives from earthquakes every year. Although all earthquakes differ in their sizes, they can occur anywhere around the globe. The demand for reducing the risk associated with earthquakes has grown every year and has led to a greater research focus on seismic design and seismic performance evaluation. Recently, a performance-based seismic engineering approach has been adopted in the earthquake engineering community. In this approach, multiple seismic performance objectives are specified explicitly: the objectives are defined with combinations of seismic hazard levels and structural and non-structural performance levels, in contrast to conventional prescriptive design approaches. Critical components of performance-based seismic design and evaluation procedures include state-of-the-art technologies involving seismic hazard analyses, robust numerical simulation frameworks, and sophisticated performance-based seismic design and assessment methodologies. Although major technologies have been developed, many challenging obstacles must be solved before they are implemented in code provisions. The Special Issue of the journal Applied Sciences "Advanced Methods for Seismic Performance Evaluation of Building Structures" aims to cover recent advances in the development of major components of seismic performance evaluation and design.

Sang Whan Han
Editor

 *applied sciences*

Article

Prediction of Structural Type for City-Scale Seismic Damage Simulation Based on Machine Learning

Zhen Xu [1], Yuan Wu [1], Ming-zhu Qi [1], Ming Zheng [1], Chen Xiong [2,*] and Xinzheng Lu [3]

[1] Beijing Key Laboratory of Urban Underground Space Engineering, School of Civil and Resource Engineering, University of Science and Technology Beijing, Beijing 100083, China; xuzhen@ustb.edu.cn (Z.X.); ivan.wuyuan@gmail.com (Y.W.); qimz1012@foxmail.com (M.-z.Q.); zhengming_521@163.com (M.Z.)
[2] Guangdong Provincial Key Laboratory of Durability for Marine Civil Engineering, Shenzhen University, Shenzhen 518060, China
[3] Key Laboratory of Civil Engineering Safety and Durability of China Education Ministry, Department of Civil Engineering, Tsinghua University, Beijing 100084, China; luxz@tsinghua.edu.cn
*   Correspondence: xiongchen@szu.edu.cn

Received: 27 January 2020; Accepted: 2 March 2020; Published: 5 March 2020

Abstract: Being the necessary data of the city-scale seismic damage simulations, structural types of buildings of a city need to be collected. To this end, a prediction method of structural types of buildings based on machine learning (ML) is proposed herein. Specifically, using the training data of 230,683 buildings in Tangshan city, China, a supervised ML solution based on a decision forest model was designed for the prediction. The scale sensitivity and regional applicability of the designed solution are discussed, respectively, and the results show that the supervised ML solution can maintain high accuracy for different scales; however, it is only suitable for cities similar to the sample city. For wide applicability for various cities, a semi-supervised ML solution was designed based on sampling investigation and self-training procedures. The downtowns of Daxing and Tongzhou districts in Beijing were selected as a case study for the designed semi-supervised ML solution. The overall prediction accuracies of structural types for Daxing and Tongzhou downtowns can reach 94.8% and 99.5%, respectively, which are acceptable for seismic damage simulations. Based on the predicted results, the distributions of seismic damage in Daxing and Tongzhou downtown were output. This study provides a smart and efficient method for obtaining structural types for a city-scale seismic damage simulation.

Keywords: machine learning; structural types; decision forest; self-training procedures; city-scale seismic damage simulation

---

1. Introduction

Generally, cities are densely organized, with many buildings and civil infrastructures. If a city is affected by a strong earthquake, many casualties and significant losses will occur. For example, the 2011 Christchurch earthquake of New Zealand caused 185 deaths and a loss of US$ 11–15 billion [1].

Earthquakes pose a serious threat for many cities in China. For instance, Tangshan, a medium-sized city in China was hit by an Ms 7.8 intraplate earthquake on 28 July 1976, which caused more than 240,000 deaths, and razed the city of Tangshan [2]. Actually, two third of cities beyond one million people in China are located in high risk areas of earthquakes (i.e., the corresponding seismic precautionary intensities of these cities are more than 6 according to the seismic design code of China [3]). For example, Beijing, the capital of China, and Taiyuan, a large city in the north of China, are both located in the area of seismic precautionary intensity 8. Therefore, the earthquake safety of these cities deserves further study.

A city-scale seismic damage simulation is important for earthquake disaster prevention and mitigation. Such a simulation can provide detailed results of potential building seismic damages, which can support decision making on urban planning for disaster prevention, seismic retrofit, earthquake emergency management, etc.

Currently, the multi-degree-of-freedom (MDOF) model proposed by Lu et al. [4–8] has been validated by actual earthquakes and successfully applied in several cities [5,8,9]. For instance, it is employed by the SimCenter project of the National Science Foundation of the United States to predict the seismic damage of 1.8 million buildings in the San Francisco Bay Area [10].

The MDOF model requires five parameters: Story area, story height, story number, construction year, and structural type. With the development of geographic information system (GIS) technology, many algorithms have been proposed to obtain the story areas and building heights automatically. For example, Li et al. proposed an algorithm for automatically detecting building footprints from very high resolution (VHR) satellite images by using a visual attention method and morphological building indices [11]. The story areas can be calculated by the polygons of building footprints. Kadhim and Mourshed proposed a shadow-overlapping algorithm for estimating building heights from VHR satellite images [12], by using graph theory and morphological fuzzy processing techniques. Note that imagery between 0.3–0.8 m/pxl is considered VRH satellite imagery in the above research. Story numbers can be calculated according to building heights and the design story heights of different building types. For example, the popular design story height is 3.0 m for residential buildings in China, according to the corresponding design code [13]. Additionally, the construction year can be quickly determined by the comparison of historical maps [14]. Currently, many cities have provided their building data for the public, e.g., the building data (e.g., footprints and construction years) of San Francisco can be downloaded by the DataSF website [15].

However, the data of structural types (e.g., masonry, frame, and shear wall structures) of the entire city are difficult to obtain, which limits the applications of city-scale seismic damage simulations based on the MDOF model. Specifically, the structural type of a building can hardly be identified by satellite image or maps; therefore, the structural type must be determined from the building's interior or by consulting relevant engineering drawings, which results in extensive manual workloads. Therefore, an efficient method to predict the structural types of building groups is necessary for a city-scale seismic damage simulation.

Machine learning (ML) [16] can be employed to predict the structural building types of a city. Using ML, some potential patterns can be obtained to perform predictions from large amounts of data. Accordingly, the potential patterns between structural types and other building inventory data (e.g., story area, story height, story number, and construction year) can be identified using ML. Consequently, the structural type of each building in a city can be predicted using such patterns.

In the field of buildings and constructions, ML is primarily adopted for automatic designing and detection [17–21]. For instance, Krijnen et al. proposed a self-learning algorithm for the automatic selections of structures based on building information models [17]. Dornaika et al. presented a generic framework that exploited recent advances in image segmentation and region descriptor extraction for the automatic and accurate detection of buildings on aerial orthophotos [18]. Yuan et al. designed a deep convolutional network with a simple structure that integrates the activation from multiple layers for pixel-wise prediction; furthermore, they trained the network to extract buildings from aerial scene images [19,20]. Bassier et al. proposed an automatic building recognition method based on ML for point cloud models created by three-dimensional laser scanners [21]. However, the existing ML-based studies on the predictions of structural types of buildings are few.

Herein, a method to predict the structural type of buildings based on ML is proposed. Specifically, using the training data of 230,683 buildings in Tangshan city, China, a supervised ML solution based on a decision forest model was designed for the prediction. The scale sensitivity and regional applicability of the designed solution are discussed, respectively, and the results show that the supervised ML solution can maintain high accuracy for different scales; however, it is only suitable for cities similar to the sample city. For wide applicability for various cities, a semi-supervised ML solution was designed based on sampling investigation and self-training procedures. The downtowns of Daxing and Tongzhou districts in Beijing were selected as a case study for the designed semi-supervised ML solution. The overall prediction accuracies of structural types for Daxing and Tongzhou downtowns can reach 94.8% and 99.5%, respectively, which are acceptable for seismic damage simulations. Based on the predicted results, the distributions of seismic damage in Daxing and Tongzhou downtown were output. This study provides a smart and efficient method for obtaining structural types for a city-scale seismic damage simulation.

2. Framework

The framework of this study is shown in Figure 1, which includes four parts: Training data, supervised ML solution, semi-supervised ML solution, and case study.

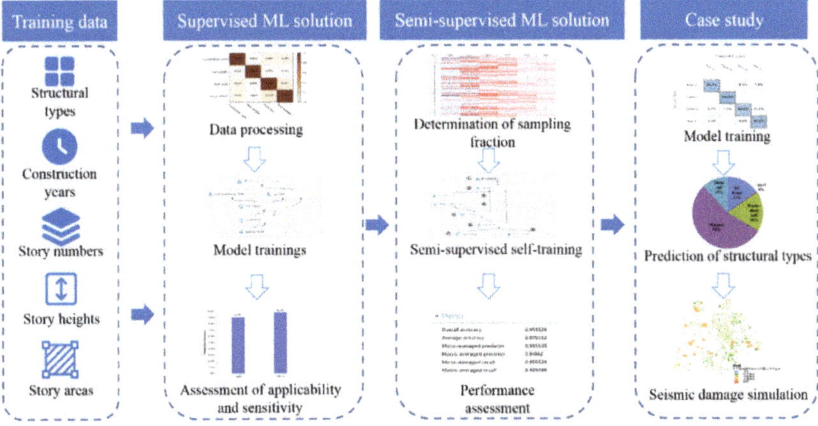

Figure 1. The framework of this study.

Training data: This part includes five attributes of a building, i.e., structural type, construction year, story number, story height, and story area. The building data of two cities in China was used for the training in this study. One city is Tangshan, which has 230,683 buildings; the other city is Taiyuan, whose downtown has 31,154 buildings. These data were provided by the department of urban planning in the local government.

Supervised ML solution: The ML models and implementation platforms suitable for the prediction of structural types are determined by comparing prediction accuracies. In addition, the scale sensitivity and regional applicability of the designed solution are discussed.

Semi-supervised ML solution: First, the sampling fraction of the building investigation is determined based on the supervised ML solution above. Subsequently, the semi-supervised self-training procedure is designed based on the sample data. Finally, the prediction performance of the semi-supervised ML solution is assessed.

Case study: The downtowns of Daxing and Tongzhou, the districts of Beijing, were selected as a case study, which has 69,180 and 34,763 buildings, respectively. The structural types of buildings in the downtowns of Daxing and Tongzhou were predicted using the designed semi-supervised ML solution, and the prediction performances were assessed based on the sample data. Furthermore, the seismic damage of Daxing and Tongzhou downtowns were simulated using the predicted structural types.

3. Supervised ML Solution

3.1. Determination of Models and Platforms

Many ML models and implementation platforms exist [22–27], and each has its own advantages and disadvantages. Therefore, appropriate models and platforms must be determined.

(1)   ML models

The purpose of this study is to predict the structural type with other building attribute data. Structural types are generally limited, e.g., masonry, frame, and shear-wall structures; therefore, the prediction of structural type is a multi-class classification problem in ML. The existing studies indicate that artificial neural network [22], decision forest [23], support vector machine (SVM) [24], and logistic regression [25] are suitable for the classification problem.

The artificial neural network model that simulates the synaptic connection of the brain comprises a large number of neurons and their interconnections; it can be used for multi-class classification problems [22]. The decision forest model is equivalent to an upgraded decision tree model. The decision forest [23] is composed of many decision trees, and each decision tree is independent. For classification problems, the prediction result with the highest accuracy in all the decision trees will be selected as the result of the decision forest. Logistic regression is the appropriate regression analysis to conduct when the dependent variable is binary or multi-class, because it can describe data and explain the relationship between one dependent variable and one or more independent variables. The SVM method is mainly used to segregate the two classes. However, the prediction of structural types in this study is a multi-class problem. Therefore, except SVM model, the artificial neural network, decision forest and logistic regression models were adopted in this study. The prediction results of these three models can be compared with each other, and more accurate results will be applied in the city-scale seismic damage simulation.

(2)   Implementation platforms

Currently, many implementation platforms exist for ML, e.g., BigML [26], Microsoft Azure (hereinafter referred to as Azure) [27], Google's TensorFlow [28], and Amazon Machine Learning [29]. Azure has integrated lots of the existing ML models, e.g., the artificial neural network, decision forest models, and logistic regression models, and it can be freely employed for a long time. Therefore, Azure was adopted in this study.

3.2. Data Processing

The data (i.e., building footprints, construction years, story heights, story areas and story numbers) of 230,683 building in Tangshan were provided by the department of urban planning in the local government of Tangshan city.

First, the department has validated the data through the extensive surveying and mapping jobs; thus, these data can be considered to be cleaned before the training.

Subsequently, the correlation matrix of the building data was calculated to evaluate possible dependencies. Taking Tangshan city, for example, the correlation matrix of the building data is demonstrated in Figure 2. It can be observed that the building data has weak dependencies and can be used as the input data for predicting structural types of buildings.

Finally, the building data have been normalized by using the min-max normalization method, because building attributes are generally concentrated within a certain range. For example, the story numbers for most buildings in Tangshan are 1–6. By the min-max normalization, the effect on the prediction caused by different scales of data can be avoided.

Note that only four types of building data are used to predict structural types in this study, thus, the building data need to be carefully checked to avoid incorrect or missing data. Actually, the purpose of the prediction of structural types in this study is to support the urban planning for earthquake preparedness, hence, the data provider of this study is the department of urban planning of a city, and they can guarantee the accuracy of the provided data.

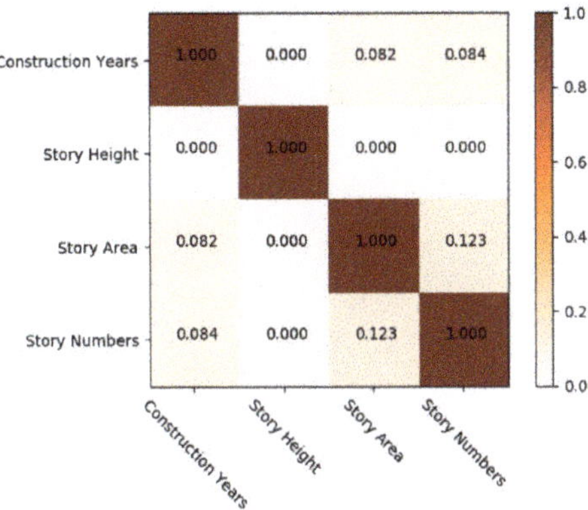

Figure 2. Correlation matrix of the building data.

3.3. Model Training

The data of 230,683 buildings in Tangshan were used to train the artificial neural network, decision forest and logistic regression models. Azure packages each operation as a component that can be defined and organized through visual programming. The prediction solution can be created efficiently using components. Using the decision forest model as an example, the supervised ML solution was designed using the components, as shown in Figure 3.

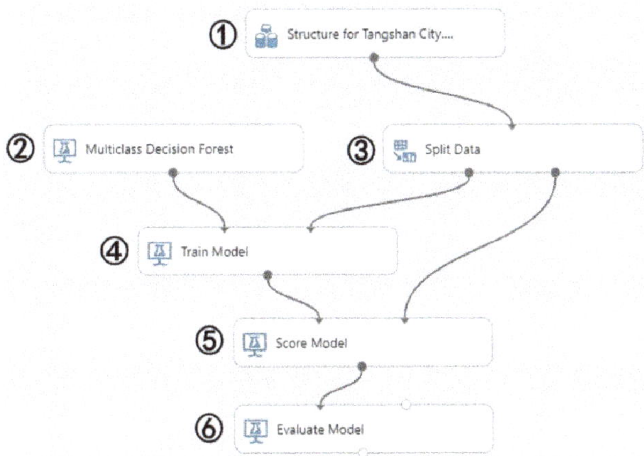

Figure 3. Designed supervised machine learning (ML) solution.

The components in Figure 3 are specified as follows:

Component 1 (Select Data): Select the uploaded source data, i.e., Tangshan data.
Component 2 (Select Model): Select "Multi-class Decision Forest" model in Azure for the prediction.
Component 3 (Split Data): Split the source data into training data (i.e., 80% of the source data) and assessment data (i.e., the remaining 20% of the source data).

Component 4 (Train Model): Train the prediction model using the training data.
Component 5 (Score Model): Score the prediction results using the assessment data.
Component 6 (Evaluate Model): Evaluate the accuracy of the prediction model.

Using the designed supervised ML solution above, the overall accuracy for predicting structural type in Tangshan reaches 98.3%, as shown in Figure 4a. If the artificial neural network and logistic regression models are adopted, the corresponding overall accuracies will be 98.0% and 97.0%, as shown in Figure 4b,c. Therefore, the designed supervised ML solution can predict the structural type of a city with high accuracy.

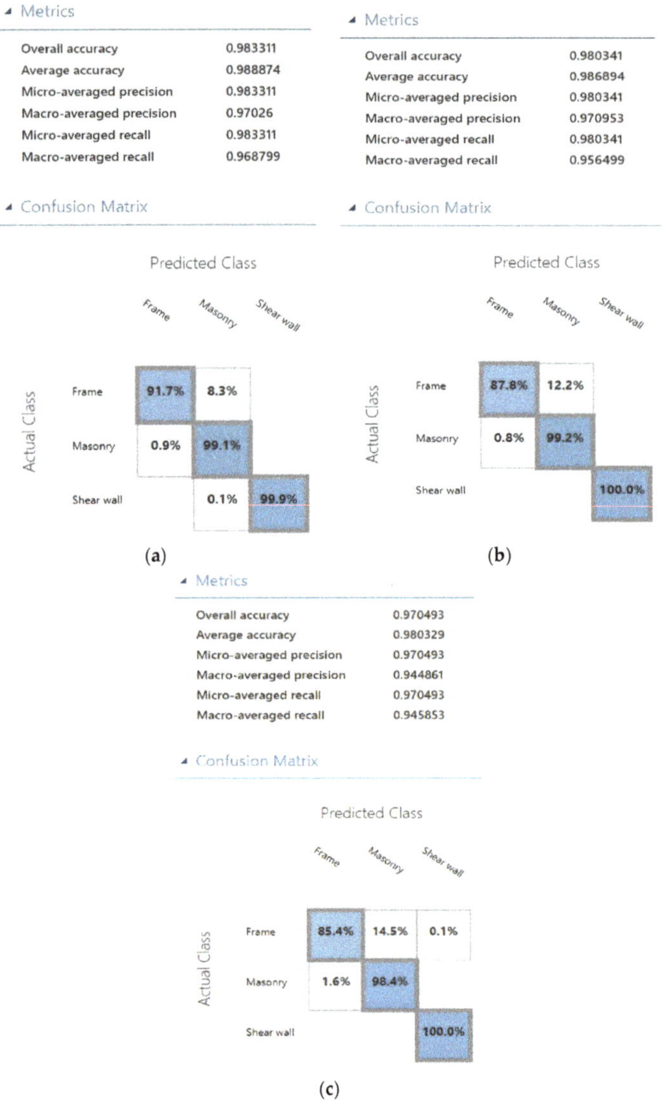

Figure 4. Prediction accuracies and confusion matrices for the designed supervised ML solution using (**a**) multi-class decision forest model, (**b**) multi-class neural network model and (**c**) logistic regression model.

Besides, according to the precisions and recalls in Figure 4, the macro and micro F1 scores of the above three predictions can be calculated to evaluate the performances of different ML models further, as shown in Table 1. The F1 score is simply a way to combine the performance metrics of precision and recall. According to Table 1, the macro and micro F1 scores of the decision forest model is the highest, while that of the logistic regression model is the lowest. In addition, the decision forest model also has the highest accuracy, but the logistic regression has the lowest accuracy. Therefore, although the artificial neural network and logistic regression models also have high accuracy, the decision forest model is recommended for the prediction of structural types in this study, due to the best performance in the above three models.

Table 1. The macro and micro F1 scores of the predictions in Tangshan.

ML Model	Macro F1 Score	Micro F1 Score
Decision forest	96.9%	98.3%
Artificial neural network	96.3%	98.0%
Logistic regression	94.6%	97.0%

3.4. Scale Sensitivity Assessment

In actual application scenarios, the scales of the predicted buildings are uncertain. To assess the scale sensitivity of the prediction model, different scales of buildings were adopted to perform the predictions. In these predictions, the building data were randomly selected from all buildings in Tangshan city. The prediction results are shown in Figure 5.

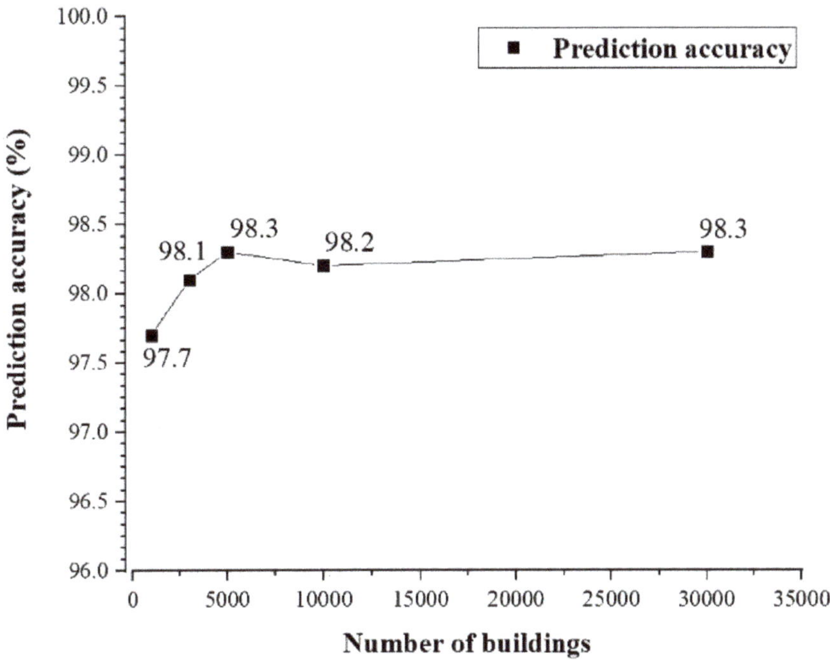

Figure 5. Prediction accuracies for different building scales.

When the building scales are 1000, 3000, 5000, 10,000, and 30,000, the corresponding prediction accuracies are 97.7%, 98.1%, 98.3%, and 98.2%, 98.3% respectively. The results show that the building scales have no significant effect on the prediction accuracy. Therefore, the designed supervised ML solution can accurately predict the structural type for different building scales.

3.5. Regional Applicability Assessment

The structural types of buildings may differ from different regions. Therefore, if the prediction model is trained with the data of a city, the prediction accuracy may decrease when it is used in other cities.

To assess the effects of different regions on the prediction accuracy, the model trained by Tangshan city was used to predict the structural types of 31,154 buildings in the downtown of Taiyuan city, China. Note that the building data of Taiyuan were provided by the department of urban planning in the local government. As demonstrated in Figure 6, the prediction accuracy of Taiyuan is 90.6% using the sample data in Tangshan, while that of Tangshan is 98.3% using the same data.

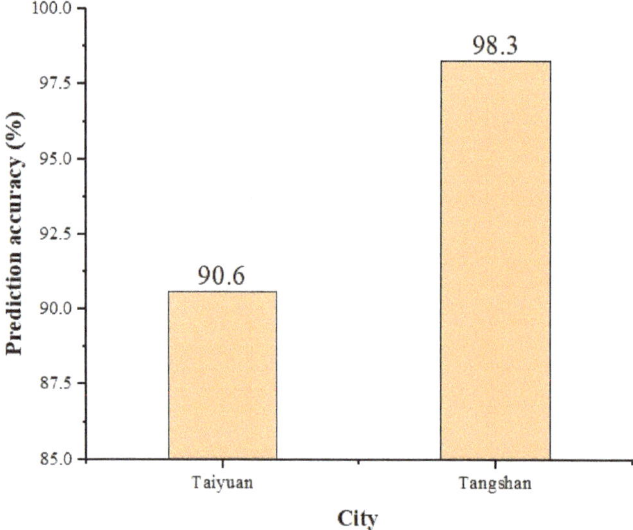

Figure 6. Prediction accuracies for different cities.

It is noted that Taiyuan and Tangshan are typical northern cities in China; therefore, the two cities are similar. However, if a southern city is used, the prediction accuracy may be not very high. Therefore, the designed supervised ML solution is recommended to be applied for cities similar to the sample city. Furthermore, a semi-supervised prediction model is designed for better regional applicability.

4. Semi-Supervised ML Solution

4.1. Determination of Sampling Fraction

The designed semi-supervised ML solution is based on the building sampling investigation in the predicted city. In detail, the sampling building is randomly selected, and then the structural types of sampling buildings are determined by consulting the relevant engineering drawings from the urban archive administration, so that the sampled structural types are accurate. The building data obtained by the sampling investigation were used to predict the structural type of all buildings in the city. In the sampling investigation, deficient building data may cause an inaccurate prediction result, while excessive building data incur extensive manual work; therefore, the number of buildings to be investigated (i.e., sampling fraction) is a critical question.

According to the existing building data of the two cities (i.e., Taiyuan and Tangshan), the predictions were performed based on different sampling fractions to determine the optimal fraction. In detail, six sampling fractions (i.e., 0.05%, 0.1%, 0.5%, 1%, 5%, and 10%) were performed for the prediction of two cities. The predictions employ the decision forest model. The sample data were used to train the model, while all the remaining building data in the city except the sample data were used for assessing the accuracy of the prediction.

For Taiyuan and Tangshan, the data were randomly sampled 50 times at each sampling fraction, and the corresponding 50 times prediction were performed according to the sampled data. The accuracies of the predictions are shown in Figures 7 and 8. The curves of prediction accuracies and variance with the sampling fractions are illustrated in Figures 9 and 10, respectively.

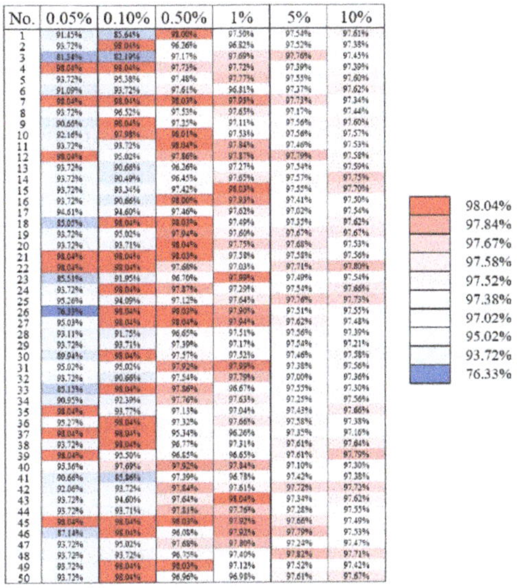

No.	0.05%	0.10%	0.50%	1%	5%	10%
1	91.15%	85.64%	98.00%	97.50%	97.54%	97.61%
2	93.72%	98.04%	96.26%	96.32%	97.52%	97.38%
3	81.54%	82.19%	97.17%	97.69%	97.76%	97.45%
4	98.04%	98.04%	97.73%	97.72%	97.39%	97.39%
5	93.72%	95.38%	97.48%	97.77%	97.55%	97.60%
6	91.09%	93.72%	97.61%	96.81%	97.37%	97.62%
7	98.04%	98.04%	98.03%	97.93%	97.73%	97.34%
8	93.72%	96.52%	97.53%	97.65%	97.17%	97.44%
9	90.66%	98.04%	97.35%	97.11%	97.56%	97.60%
10	92.16%	97.98%	96.01%	97.53%	97.56%	97.57%
11	93.72%	93.72%	98.04%	97.84%	97.46%	97.53%
12	98.04%	95.02%	97.86%	97.87%	97.79%	97.58%
13	93.72%	90.66%	96.26%	97.27%	97.54%	97.59%
14	93.72%	90.49%	96.45%	97.65%	97.57%	97.75%
15	93.72%	93.34%	97.42%	98.03%	97.55%	97.70%
16	93.72%	90.66%	98.00%	97.93%	97.41%	97.50%
17	94.61%	94.60%	97.46%	97.62%	97.02%	97.54%
18	85.03%	98.04%	98.03%	97.49%	97.35%	97.62%
19	93.72%	95.02%	97.94%	97.60%	97.67%	97.67%
20	93.72%	93.71%	98.04%	97.75%	97.68%	97.53%
21	98.04%	98.04%	98.03%	97.58%	97.58%	97.56%
22	98.04%	98.04%	97.68%	97.03%	97.71%	97.80%
23	85.51%	91.95%	96.70%	97.99%	97.49%	97.54%
24	93.72%	98.04%	97.87%	97.29%	97.54%	97.66%
25	95.26%	94.09%	97.12%	97.64%	97.76%	97.73%
26	76.33%	98.04%	98.03%	97.90%	97.51%	97.55%
27	95.03%	98.04%	98.04%	97.94%	97.62%	97.48%
28	93.11%	91.75%	96.65%	97.51%	97.56%	97.39%
29	93.72%	93.71%	97.39%	97.17%	97.54%	97.21%
30	89.94%	98.04%	97.57%	97.52%	97.46%	97.58%
31	95.02%	95.02%	97.92%	97.99%	97.38%	97.56%
32	93.72%	90.66%	97.54%	97.79%	97.00%	97.36%
33	85.15%	98.04%	97.86%	96.67%	97.55%	97.30%
34	90.95%	92.39%	97.76%	97.63%	97.25%	97.56%
35	98.04%	93.77%	97.13%	97.04%	97.43%	97.66%
36	95.27%	98.04%	97.32%	97.66%	97.58%	97.38%
37	98.04%	98.04%	95.34%	96.26%	97.35%	97.16%
38	93.72%	98.04%	96.77%	97.31%	97.61%	97.64%
39	98.04%	92.50%	96.65%	96.65%	97.61%	97.79%
40	95.36%	97.69%	97.92%	97.84%	97.10%	97.30%
41	90.66%	85.86%	97.39%	96.78%	97.42%	97.38%
42	92.06%	93.72%	97.84%	97.61%	97.72%	97.72%
43	93.72%	94.60%	97.64%	98.04%	97.34%	97.62%
44	93.72%	93.71%	97.81%	97.76%	97.28%	97.55%
45	98.04%	98.04%	98.03%	97.92%	97.66%	97.49%
46	87.14%	98.04%	96.08%	97.92%	97.79%	97.53%
47	93.72%	95.02%	97.68%	97.80%	97.24%	97.47%
48	93.72%	93.72%	96.75%	97.40%	97.82%	97.71%
49	93.72%	98.04%	98.03%	97.12%	97.52%	97.42%
50	93.72%	98.04%	96.96%	96.98%	97.61%	97.67%

Legend: 98.04% / 97.84% / 97.67% / 97.58% / 97.52% / 97.38% / 97.02% / 95.02% / 93.72% / 76.33%

Figure 7. Distribution of prediction accuracies with different sampling fractions in Taiyuan.

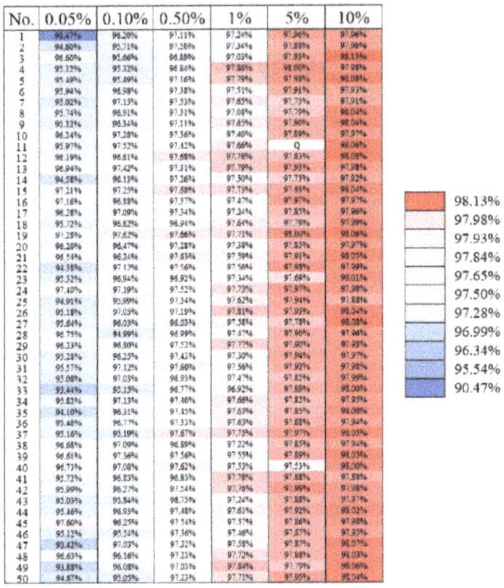

No.	0.05%	0.10%	0.50%	1%	5%	10%
1	90.47%	96.20%	97.11%	97.24%	97.96%	97.96%
2	94.60%	95.71%	97.20%	97.34%	97.88%	97.96%
3	96.60%	95.66%	96.89%	97.03%	97.93%	98.13%
4	95.32%	95.32%	96.84%	97.86%	98.00%	97.98%
5	95.49%	95.49%	97.16%	97.79%	97.98%	98.08%
6	95.94%	96.98%	97.38%	97.51%	97.91%	97.93%
7	95.02%	97.15%	97.53%	97.65%	97.75%	97.91%
8	95.74%	96.91%	97.31%	97.08%	97.79%	98.04%
9	95.33%	96.34%	97.11%	97.65%	97.90%	98.04%
10	96.34%	97.28%	97.36%	97.40%	97.89%	97.97%
11	95.97%	97.52%	97.42%	97.66%	Q	98.06%
12	98.19%	96.61%	97.68%	97.78%	97.83%	98.08%
13	96.94%	97.42%	97.31%	97.79%	97.95%	97.98%
14	94.58%	96.13%	97.26%	97.50%	97.73%	97.92%
15	97.21%	97.25%	97.68%	97.73%	97.93%	98.04%
16	97.16%	96.88%	97.37%	97.47%	97.97%	97.97%
17	96.28%	97.09%	97.34%	97.34%	97.85%	97.96%
18	95.72%	96.82%	96.94%	97.64%	97.79%	97.99%
19	97.28%	97.62%	97.66%	97.31%	98.00%	98.08%
20	96.20%	96.47%	97.28%	97.38%	97.83%	97.97%
21	96.54%	96.24%	97.63%	97.59%	97.91%	98.05%
22	94.38%	97.12%	97.56%	97.56%	97.98%	97.99%
23	95.32%	96.94%	96.92%	97.34%	97.69%	98.01%
24	97.40%	97.19%	97.52%	97.73%	97.97%	97.98%
25	94.91%	95.99%	97.34%	97.62%	97.94%	97.88%
26	95.18%	97.03%	97.19%	97.81%	97.93%	98.04%
27	95.64%	96.03%	96.03%	97.58%	97.78%	98.08%
28	95.75%	94.99%	96.99%	97.47%	97.90%	97.96%
29	96.23%	96.80%	97.52%	97.77%	97.90%	97.93%
30	95.28%	96.25%	97.43%	97.30%	97.94%	97.97%
31	95.57%	97.12%	97.60%	97.56%	97.92%	97.98%
32	95.08%	97.03%	96.93%	97.47%	97.82%	97.99%
33	93.44%	95.15%	96.77%	96.92%	97.89%	98.00%
34	95.82%	97.13%	97.10%	97.66%	97.83%	97.95%
35	94.10%	96.31%	97.45%	97.63%	97.85%	98.00%
36	95.48%	96.77%	97.33%	97.63%	97.88%	97.94%
37	95.16%	95.19%	97.67%	97.75%	97.97%	98.03%
38	96.68%	97.09%	96.89%	97.22%	97.85%	97.94%
39	96.61%	97.36%	97.56%	97.55%	97.89%	98.05%
40	96.73%	97.08%	97.62%	97.53%	97.53%	98.20%
41	95.72%	96.83%	96.83%	97.78%	97.88%	97.88%
42	95.99%	96.27%	97.54%	97.76%	97.99%	97.98%
43	95.03%	95.84%	96.75%	97.24%	97.88%	97.97%
44	95.46%	96.93%	97.48%	97.61%	97.92%	98.02%
45	97.60%	96.25%	97.54%	97.57%	97.86%	97.98%
46	95.12%	95.54%	97.36%	97.46%	97.87%	97.95%
47	93.42%	97.03%	97.22%	97.58%	97.87%	98.02%
48	96.63%	96.16%	97.23%	97.72%	97.88%	98.03%
49	93.88%	96.08%	97.03%	97.84%	97.79%	98.06%
50	94.87%	95.03%	97.23%	97.71%	97.95%	98.04%

Legend: 98.13% / 97.98% / 97.93% / 97.84% / 97.65% / 97.50% / 97.28% / 96.99% / 96.34% / 95.54% / 90.47%

Figure 8. Distribution of prediction accuracies with different sampling fractions in Tangshan.

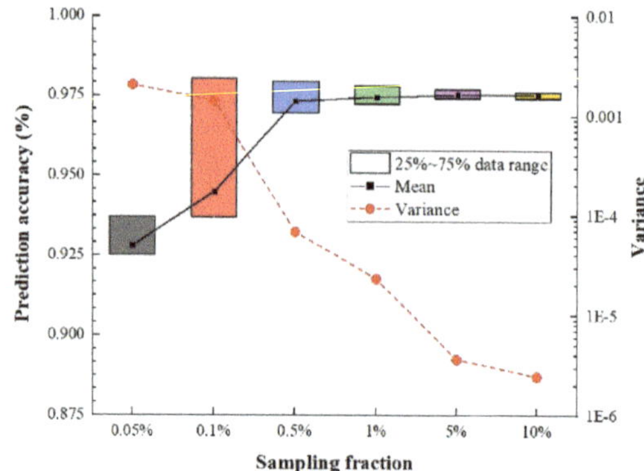

Figure 9. Prediction accuracy and variance of different sampling fractions in Taiyuan.

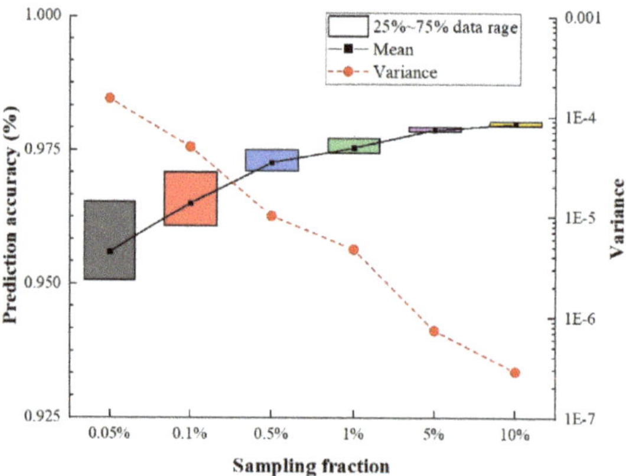

Figure 10. Prediction accuracy and variance of different sampling fractions in Tangshan.

According to the prediction results of Taiyuan and Tangshan (shown in Figures 9 and 10, respectively), when the sampling fraction is greater than 1%, the prediction is highly accurate (i.e., above 97% in two cases) and the variance of the prediction accuracies decreases. Therefore, the sampling fraction of 1% is recommended for predicting the structural type of a city.

4.2. Semi-Supervised ML Solution

The designed semi-supervised solution is shown in Figure 11. First, the sample data are obtained by the building investigation. The aforementioned prediction results indicate that when the sampling fraction of the building investigation is 1%, the prediction will be highly accurate. In this study, the sampling fraction of 3% was adopted. In detail, 1% of the sample data was used for training the model, while 2% of sample data for assessing the accuracy of the prediction. Subsequently, the designed supervised ML solution indicated that the decision forest model has the best performance for the prediction of structural type; therefore, the decision forest model was selected. Finally, a self-training process [30–32] was performed iteratively until the prediction accuracy was accepted. Specifically, the ML model was trained by the sample data, and then the prediction of the trained model was scored and evaluated, separately. If the prediction accuracy of the trained model is accepted, then the training process will end; otherwise, building data with high accuracies will be selected, and these data will be used for the next training to obtain better prediction results. Such an iterative training process is defined as a self-training process.

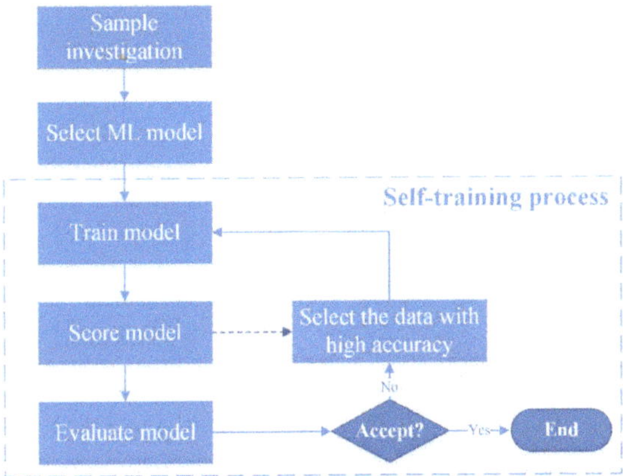

Figure 11. Designed semi-supervised ML solution.

Using Tangshan as an example, the designed semi-supervised training process is demonstrated as follows.

(1) First self-training

The process of first self-training can be implemented using the components in Azure, as shown in Figure 12.

Component 1 (Data Set): Select the sample data from the building investigation, i.e., 3% of buildings in Tangshan.

Component 2 (Split Data): 1/3 of the sample data are used for training the model (i.e., Component 5), while the remaining 2/3 of the data are used for scoring the model (i.e., Component 6).

Component 3 (Convert to CSV): Export the training data to CSV format for the subsequent training.

Component 4 (Multi-class Decision Forest): Select the multi-class decision forest model for prediction.

Component 5 (Train Model): Train the selected ML model using the training data.

Component 6 (Score Model): Score the accuracies of the prediction results with the assessment data, as shown in Figure 13. By doing this, the building data with high accuracies will be identified. In this study, the building data ranked top 1% in the scored probabilities will be selected for the next training.

Component 7 (Evaluate Model): Evaluate the performance of the prediction model and output the evaluation results. As shown in Figure 14, the overall prediction accuracy of the first self-training reaches 95.5%. However, the accuracy of the frame structure is only 81.7%, which is not acceptable; therefore, a second self-training is required.

Component 8 (Convert to CSV): Convert the building data with high accuracies (see Component 6) to CSV format such that these data can be used for the second self-training.

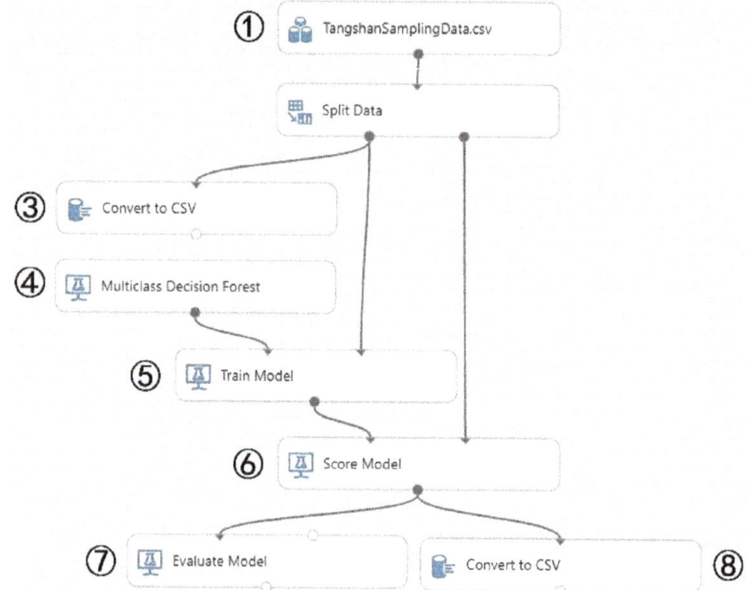

Figure 12. Process of first self-training.

ID	Scored Probabilities for Class ""Masonry""	Scored Probabilities for Class ""Frame""	Scored Probabilities for Class ""Shear Wall""	Maximum Probability	Scored Labels
4212	0.007626	0.004909	0.999604	0.999604	Shear Wall
4211	0.007608	0.005033	0.999594	0.999594	Shear Wall
4844	0.004921	0.998857	0.000024	0.998857	Frame
4122	0.010772	0.013292	0.998089	0.998089	Shear Wall
4197	0.010771	0.013298	0.998089	0.998089	Shear Wall
4206	0.010771	0.013299	0.998088	0.998088	Shear Wall
4207	0.010771	0.0133	0.998088	0.998088	Shear Wall
4196	0.010771	0.013305	0.998088	0.998088	Shear Wall
2756	0.01077	0.013309	0.998087	0.998087	Shear Wall
3870	0.01077	0.013309	0.998087	0.998087	Shear Wall
2754	0.01077	0.013318	0.998086	0.998086	Shear Wall
3891	0.010769	0.01332	0.998086	0.998086	Shear Wall

Figure 13. Part of the scored building data.

(2)   Second self-training

Second self-training was implemented using components in Azure, as shown in Figure 15. However, the training data used were different from those of the first self-training. The original training data (the CSV file in Component 3 in Figure 12) and the identified building data with high accuracy (the CSV file in Component 8 in Figure 12) were integrated as the training data for the second self-training. It is noted that the assessment data (i.e., 2/3 of the sample data) were the same for all self-trainings.

◢ Metrics

Overall accuracy	0.955528
Average accuracy	0.970352
Micro-averaged precision	0.955528
Macro-averaged precision	0.91662
Micro-averaged recall	0.955528
Macro-averaged recall	0.929388

◢ Confusion Matrix

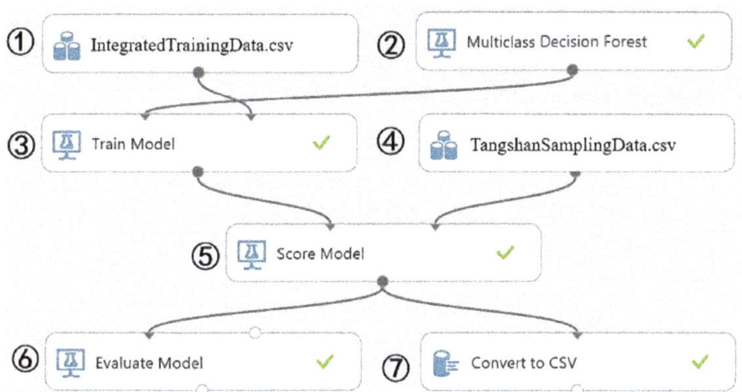

Figure 14. Prediction accuracies and confusion matrix for the first self-training scenario.

Figure 15. Process of second self-training.

Similar to the first self-training, the model was trained and evaluated. As shown in Figure 16, the overall accuracy rate is above 95.1%, which is similar to that of the first self-training (95.5%). In particular, the prediction accuracy of the frame structure increased from 81.7% to 86.9%, which is an improvement. Similarly, the building data with high accuracies were converted to CSV format for the next self-training.

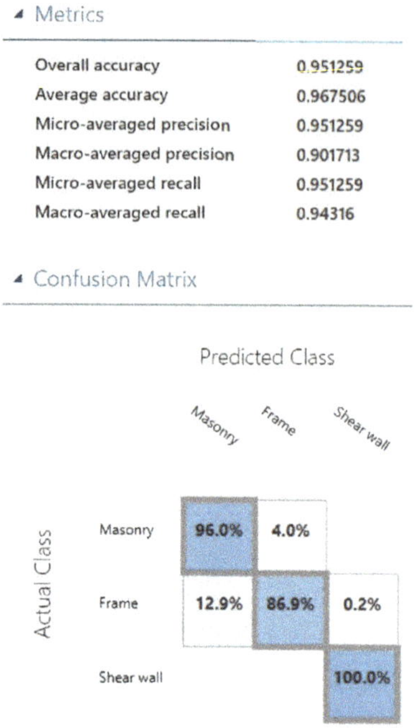

◢ Metrics

Overall accuracy	0.951259
Average accuracy	0.967506
Micro-averaged precision	0.951259
Macro-averaged precision	0.901713
Micro-averaged recall	0.951259
Macro-averaged recall	0.94316

◢ Confusion Matrix

Predicted Class

Actual Class	Masonry	Frame	Shear wall
Masonry	96.0%	4.0%	
Frame	12.9%	86.9%	0.2%
Shear wall			100.0%

Figure 16. Prediction accuracies and confusion matrix for the second self-training scenario.

(3) Third self-training

After the third self-training, the prediction results are as shown in Figure 17. The overall prediction accuracy of the model is 95.2%. In detail, the prediction accuracies of the masonry, frame, and shear wall structures are 96.1%, 87.0%, and 100.0%, respectively. The accuracies of the overall prediction and each type of structure will not increase significantly compared with those of the last self-training. In addition, the corresponding precision and recall are also very high, as shown in Figure 17. Therefore, additional self-trainings are not required.

The predicted structural type of all the buildings by the third self-training was compared with the real data in Tangshan, and the error is shown in Table 2. The results indicate that the designed semi-supervised ML solution can achieve a high prediction accuracy even when using 1% of all the building data. Both the other attribute data of buildings and the structural types of the sampling buildings are accurate, which guarantees the prediction accuracy using the designed ML solution. Furthermore, the self-learning process can improve the prediction accuracy. Therefore, the prediction results in Table 2 is exact compared with the real structural types of buildings in Tangshan.

Table 2. Comparison between the real data and prediction result in Tangshan.

Structure Type	Real Data	Prediction Results	Error
Masonry	87.07%	83.68%	3.40%
Frame	10.78%	14.18%	−3.40%
Shear wall	2.14%	2.14%	0.00%

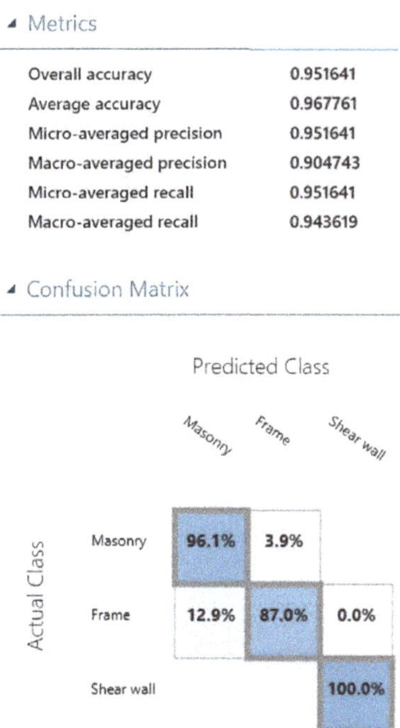

Figure 17. Prediction accuracies and confusion matrix for the third self-training scenario.

5. Case Study

5.1. Introduction of Case Study

Daxing and Tongzhou are two districts in Beijing, as shown in Figure 18. The earthquake risk of these two districts are very high (i.e., the precautionary seismic intensity is 8). In this intensity, the peak ground acceleration corresponding to the service level earthquake whose probability of exceedance in 50 years is 63.3% reaches 0.2 g [3]. Seismic damage simulation of these two districts will provide the decision-making references for their urban planning on earthquake preparedness. However, the structural types of buildings are unavailable for these two districts. Therefore, the designed ML solution will be applied in the downtowns of Daxing and Tongzhou for obtaining the data of structural types.

5.2. Structural Type Prediction for Daxing Downtown

The department of urban planning in the local government has provided the GIS data of building footprints of Daxing downtown, which includes the attributes of story area, construction years, story heights and story numbers. According to the existing GIS data, 69,180 buildings exist in Daxing downtown. In detail, the ratios of buildings constructed before 1989, from 1989 to 2001, and after 2001 are 62%, 32%, and 6%, respectively. The distribution of construction year is shown in Figure 19. It is noted here that the codes for the seismic design of buildings were updated in 1989 and 2001; therefore, buildings that were constructed later exhibit higher anti-seismic capabilities. Additionally, most buildings in this area are low rise, e.g., the number of buildings within two floors constitutes 91% of the total number. The distribution of story number is shown in Figure 20.

Figure 18. The locations of Daxing and Tongzhou districts in Beijing.

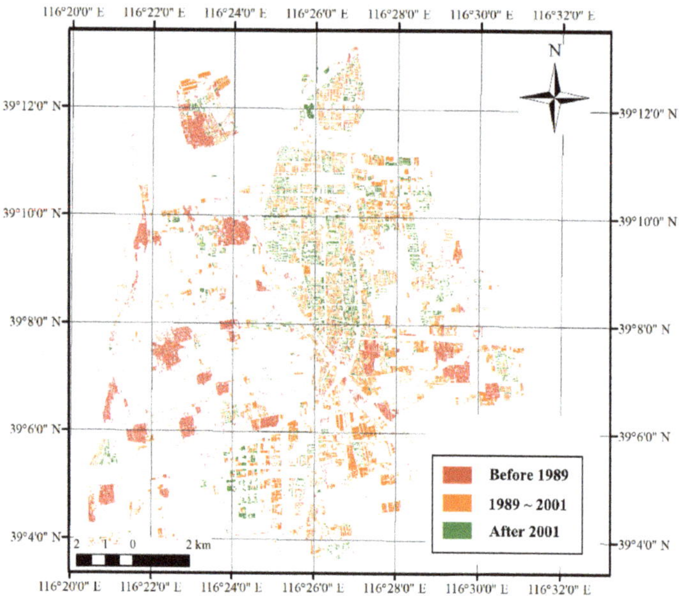

Figure 19. Building distribution with different construction years.

No structural type exists in the GIS data of Daxing downtown; therefore, the designed semi-supervised ML solution in this study was used to predict the structural type of buildings.

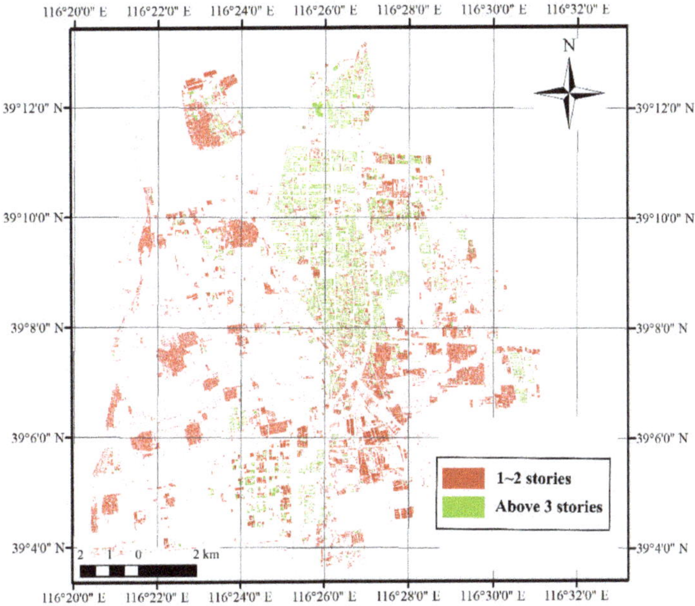

Figure 20. Building distribution with different story numbers.

(1) Building sampling investigation

In this case study, 3% of the total buildings in Daxing downtown was investigated to obtain the complete attribution data of the buildings. According to the recommended sampling fraction in this study, 1% of the total buildings was used for the training model, while the remaining 2% for assessing the model.

As mentioned previously, 69,180 buildings exist in Daxing downtown; therefore, 2075 buildings (i.e., 3% of total buildings) were randomly investigated. The distribution of structural type in the investigated buildings is shown in Table 3.

Table 3. Structure types and numbers of investigated buildings.

Masonry	Frame	Shear Wall	Light Steel	Total
385	238	761	691	2075

(2) Training model

The sample data of 2075 buildings were uploaded to the Azure platform. Using the designed semi-supervised ML solution, three self-trainings were performed, and the optimal training results are shown in Figure 21.

As shown in Figure 21, the overall accuracy of the prediction results is 94.8%. The accuracy of the frame structure is 84.0%, whereas, those of other structures are more than 92.2%. In addition, the F1 score can be calculated using precision and recall in Figure 21. In detail, the macro and micro F1 scores are 94.8% and 92.9%, respectively, which are accepted. Generally, the trained model exhibits high accuracy and performance for the prediction of structural type.

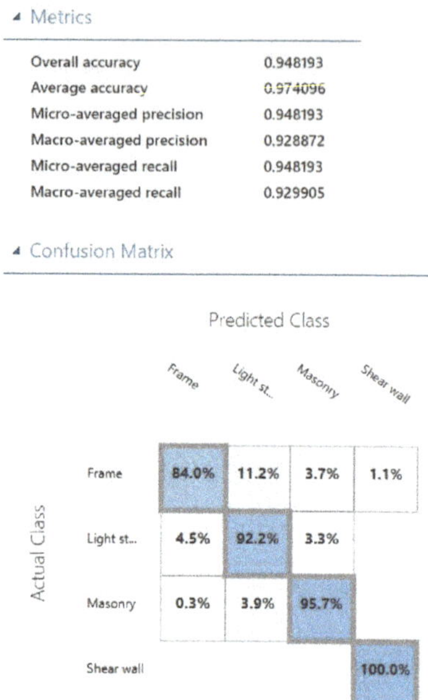

Figure 21. Prediction accuracies and confusion matrix for the case of Daxing downtown.

(3)   Prediction of structural type

Using the trained model, the structural type of all buildings in Daxing downtown were predicted. The ratios of different structural types are shown in Figure 22. It is clear that most buildings in Daxing downtown are masonry buildings, which account for 66% of the total buildings, while the frame shear wall buildings are the least, i.e., only 1% of the total buildings. The distribution of structural type in Daxing downtown is shown in Figure 23. It is noted that only 3% of all the buildings must be investigated manually when the designed semi-supervised ML solution is used. Compared with the manual investigation of all the buildings (i.e., a total of 69,180 buildings), the designed solution can save 97% of manual work and significantly improve the efficiency for obtaining the data of structural type.

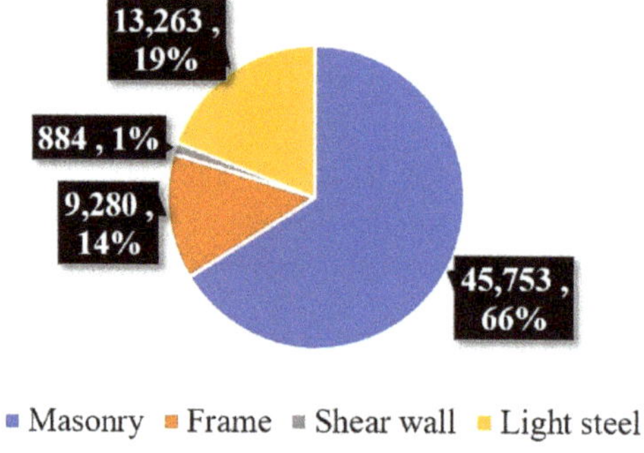

Figure 22. Ratios of structural type in Daxing downtown.

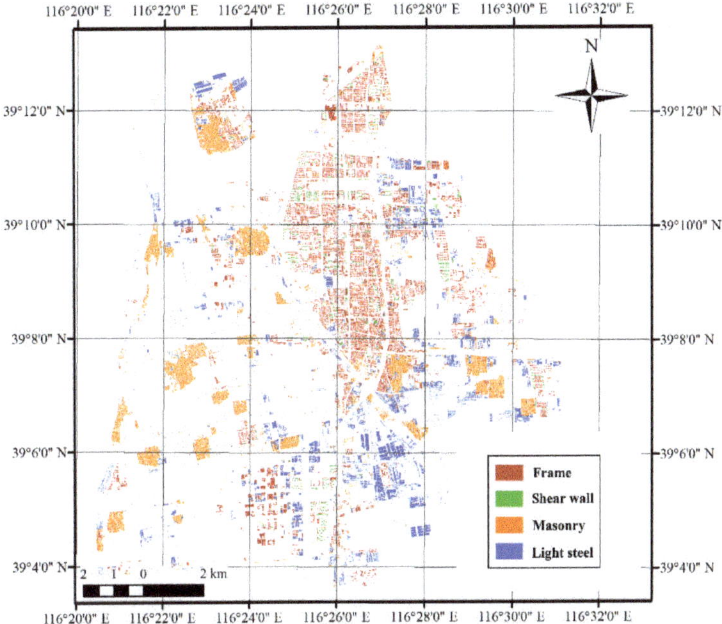

Figure 23. Distribution of building structure type in Daxing downtown.

5.3. Seismic Damage Simulation for Daxing Downtown

The Sanhe-Pinggu M 8.0 earthquake [33] occurred in 1679, which is the latest M 8.0 earthquake in the history of Beijing. The ground motion of the Sanhe-Pinggu M 8.0 earthquake was simulated [34], and the corresponding time-history accelerations are shown in Figure 24. The time-history accelerations of the Sanhe-Pinggu earthquake will be input for the seismic damage simulation of Daxing downtown.

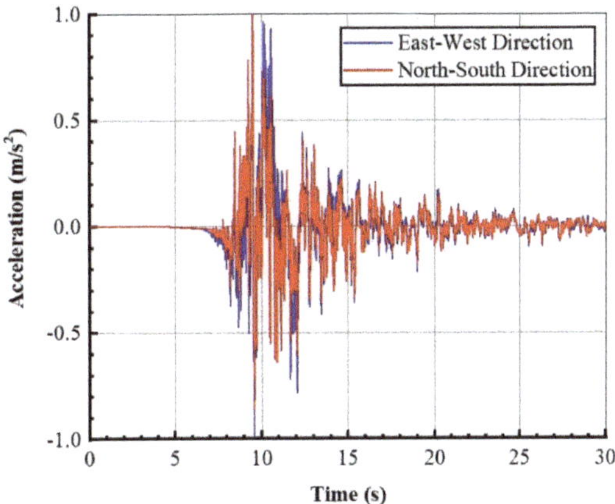

Figure 24. Time-history accelerations in the Sanhe-Pinggu earthquake.

Based on the predicted building data, the MDOF models of buildings were created, and the seismic damage of Daxing downtown was simulated based on the MDOF models and nonlinear time–history analysis. The distribution of seismic damage in Daxing downtown is shown in Figure 25. It is clear that most buildings suffer slight damages. To reveal the features of the seismic damages clearly, the damage ratios of different structural types and construction years are shown in Figure 26. As shown, buildings of masonry structure and constructed before 1989 suffer severe damages, which provides important references for seismic retrofit decisions. It is noted that the seismic damage simulation of a city is implemented based on the designed semi-supervised ML solution, which is useful for improving the seismic capability of the city.

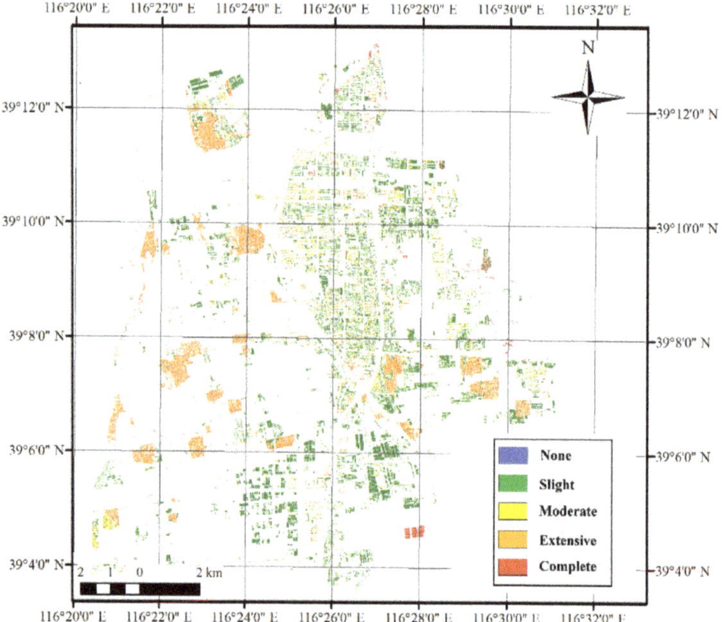

Figure 25. Distribution of seismic damage in Daxing downtown.

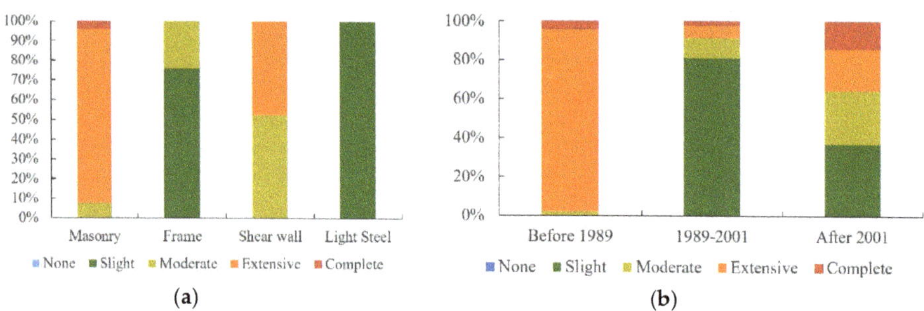

Figure 26. Seismic damage statistics by (**a**) structure type and (**b**) construction year.

5.4. Structural Type Prediction for Tongzhou Downtown

The GIS data of building footprints of Tongzhou downtown are also from the department of urban planning in the local government. According to the GIS data, Tongzhou downtown has 37,463 buildings.

By the building sampling investigation, the structural types of 3% of the total buildings in Tongzhou downtown were determined. According to the recommended sampling fraction in this study, 1% of the total buildings was used for the training model, while the remaining 2% for assessing the model.

Using the designed semi-supervised ML solution, four self-trainings were performed, and the optimal training results are shown in Figure 27. As shown in Figure 27, the overall accuracy of the prediction results is 99.5%, and the accuracy of each structure type is beyond 97.9%. Obviously, the trained model exhibits a high prediction accuracy. Besides, the corresponding precisions and recalls are beyond 99.0%, as shown in Figure 27.

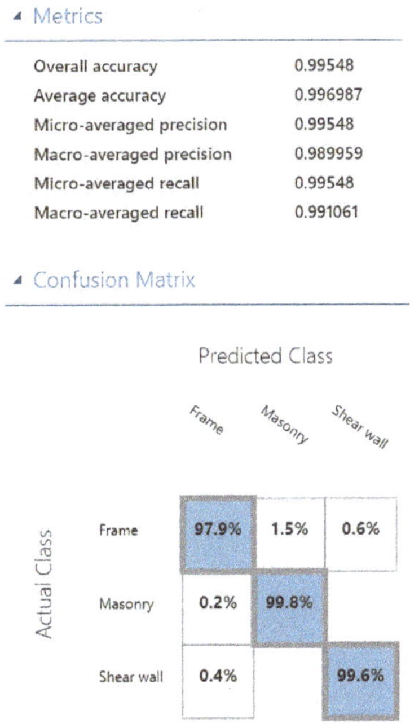

Figure 27. Prediction accuracies and confusion matrix for the case of Tongzhou downtown.

Using the trained model, the structural type of all buildings in Tongzhou downtown were predicted. The ratios of different structural types are shown in Figure 28. It is clear that most buildings in Tongzhou downtown are masonry buildings, which is similar to Daxing downtown. The distribution of structural type in Tongzhou downtown is shown in Figure 29.

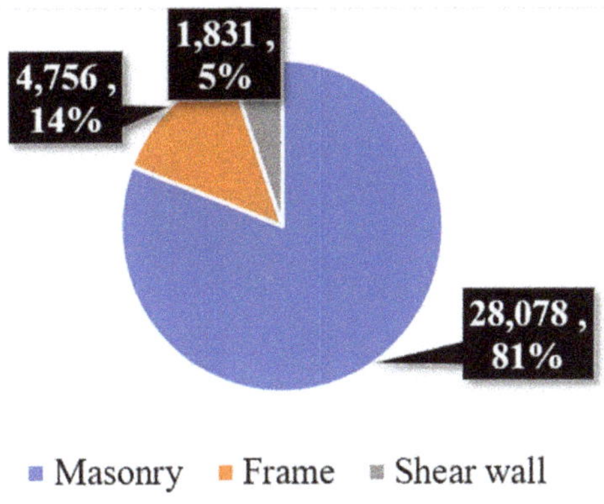

Figure 28. Ratios of structural type in Tongzhou downtown.

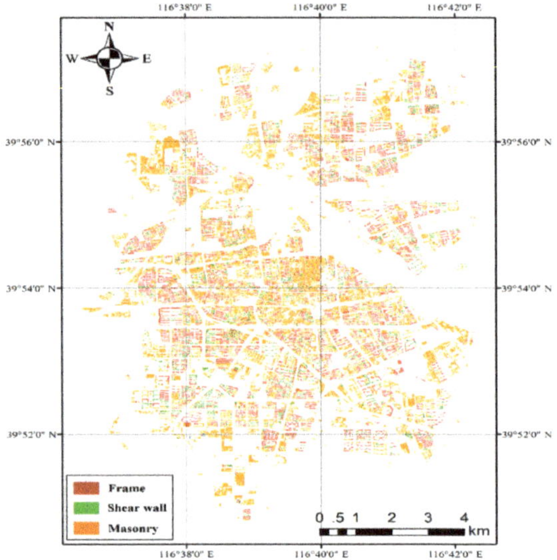

Figure 29. Distribution of structure type in Tongzhou downtown.

5.5. Seismic Damage Simulation for Tongzhou Downtown

The ground motion of the Sanhe-Pinggu M 8.0 earthquake was also used for the seismic damage simulation of Tongzhou downtown. Based on the predicted building data and the MDOF models, the seismic damage of Tongzhou downtown was simulated, as shown in Figure 30. It is clear that most buildings suffer slight and moderate damages. The simulated seismic damage will provide the decision-making references for the urban planning on earthquake preparedness (e.g., seismic retrofitting planning).

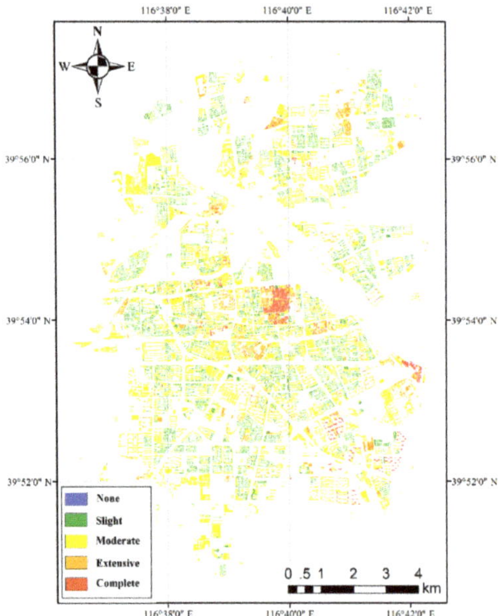

Figure 30. Distribution of seismic damage in Tongzhou downtown.

6. Conclusions

An ML-based prediction method of the structural type of buildings was proposed, and the case study of Daxing and Touzhou downtowns in Beijing were investigated. Some conclusions are drawn as follows:

(1) The prediction result of the designed supervised ML solution for Tangshan with 230,683 buildings indicated that decision forest, artificial neural network and logistic regression models exhibited high prediction accuracy. Especially, the decision forest model has the best performance and is recommended to predict structural types.

(2) The designed supervised ML solution could maintain high prediction accuracy for different building scales; however, it should be applied for cities similar to the sample city.

(3) The designed semi-supervised ML solution was applicable to different cities, based on a sampling investigation. According to the prediction with different sampling fractions, the sampling fraction of 1% is recommended. Through multiple self-trainings, the semi-supervised ML solution achieved high accuracy for predicting structural types.

(4) This study provided a smart and efficient method to predict structural type for a city-scale seismic damage simulation.

Author Contributions: Conceptualization, Z.X. and M.Z.; methodology, Y.W.; software, M.-z.Q.; validation, C.X.; resources, X.L.; writing—original draft preparation, Z.X.; writing—review and editing, M.-z.Q. All authors have read and agreed to the published version of the manuscript.

Funding: This research was funded by Scientific Research Fund of Institute of Engineering Mechanics, China Earthquake Administration (Grant No. 2019EEEVL0501), General Program of the National Natural Science Foundation of China (Grant No. 51978049) and Beijing Nova Program of Science and Technology (Grant No. Z191100001119115).

Conflicts of Interest: The authors declare no conflict of interest.

References

1. Reyners, M. Lessons from the destructive mw 6.3 Christchurch, New Zealand, earthquake. *Seismol. Res. Lett.* **2011**, *82*, 371–372. [CrossRef]
2. Li, Z.; Ni, S.; Roecker, S. Seismic imaging of source region in the 1976 M s 7.8 Tangshan earthquake sequence and its implications for the seismogenesis of intraplate earthquakes. *Bull. Seismol. Soc. Am.* **2018**, *108*, 1302–1313. [CrossRef]
3. Ministry of Housing and Urban-Rural Development. *Code for Seismic Design of Building (GB 50011-2010)*; China Architecture Industry Press: Beijing, China, 2016.
4. Lu, X.; Han, B.; Hori, M.; Xiong, C.; Xu, Z. A coarse-grained parallel approach for seismic damage simulations of urban areas based on refined models and GPU/CPU cooperative computing. *Adv. Eng. Softw.* **2014**, *70*, 90–103. [CrossRef]
5. Lu, X.; Guan, H. *Earthquake Disaster Simulation of Civil Infrastructures: From Tall Buildings to Urban Areas*; Springer and Science Press: Beijing, China, 2017.
6. Xiong, C.; Lu, X.; Guan, H.; Xu, Z. A nonlinear computational model for regional seismic simulation of tall buildings. *Bull. Earthq. Eng.* **2016**, *14*, 1047–1069. [CrossRef]
7. Xiong, C.; Lu, X.; Lin, X.; Ye, L. Parameter determination and damage assessment for THA-based regional seismic damage prediction of multi-story buildings. *J. Earthq. Eng.* **2017**, *21*, 461–485. [CrossRef]
8. Xu, Z.; Lu, X.; Guan, H.; Han, B.; Ren, A. Seismic damage simulation in urban areas based on a high-fidelity structural model and a physics engine. *Nat. Hazards* **2014**, *71*, 1679–1693. [CrossRef]
9. Xiong, C.; Lu, X.; Huang, J.; Guan, H. Multi-LOD seismic-damage simulation of urban buildings and case study in Beijing CBD. *Bull. Earthq. Eng.* **2019**, *17*, 2037–2057. [CrossRef]
10. Lu, X.; Frank, M.; Cheng, Q.L.; Xu, Z.; Zeng, X.; Stephen, M. An open-source framework for regional earthquake loss estimation using the city-scale nonlinear time-history analysis. *Earthq. Spectra* **2020**. [CrossRef]

11. Li, S.; Tang, H.; Huang, X.; Mao, T.; Niu, X. Automated detection of buildings from heterogeneous VHR satellite images for rapid response to natural disasters. *Remote Sens.* **2017**, *9*, 1177. [CrossRef]

12. Kadhim, N.; Mourshed, M. A shadow-overlapping algorithm for estimating building heights from VHR satellite images. *IEEE Geosci. Remote Sens. Soc.* **2017**, *15*, 8–12. [CrossRef]

13. Ministry of Housing and Urban-Rural Development. *Design Code for Residential Buildings (GB 50096-2011)*; China Architecture Industry Press: Beijing, China, 2011.

14. Tian, J.; Cui, S.; Reinartz, P. Building change detection based on satellite stereo imagery and digital surface models. *IEEE Geosci. Remote Sens. Soc.* **2013**, *52*, 406–417. [CrossRef]

15. DataSF. Available online: https://datasf.org/opendata/ (accessed on 22 February 2020).

16. Bishop, C.M. *Pattern Recognition and Machine Learning*; Springer: New York, NY, USA, 2006.

17. Krijnen, T.; Tamke, M. Assessing Implicit Knowledge in BIM Models with Machine Learning. In *Modelling Behaviour*; Springer: Den Dolech, The Netherlands, 2015; pp. 397–406.

18. Dornaika, F.; Moujahid, A.; El Merabet, Y.; Ruichek, Y. Building detection from orthophotos using a machine learning approach: An empirical study on image segmentation and descriptors. *Expert Syst. Appl.* **2016**, *58*, 130–142. [CrossRef]

19. Yuan, J.; Cheriyadat, A.M. Combining maps and street level images for building height and facade estimation. In Proceedings of the 2nd ACM SIGSPATIAL Workshop on Smart Cities and Urban Analytics, Burlingame, CA, USA, 31 October 2016; p. 8.

20. Yuan, J. Learning building extraction in aerial scenes with convolutional networks. *IEEE Trans. Pattern Anal. Mach. Intell.* **2017**, *40*, 2793–2798. [CrossRef] [PubMed]

21. Bassier, M.; Vergauwen, M.; Van Genechten, B. Automated classification of heritage buildings for as-built BIM using machine learning techniques. *ISPRS Ann. Photogramm. Remote Sens. Spat. Inf. Sci.* **2017**, *IV-2/W2*, 25–30. [CrossRef]

22. Dreiseitl, S.; Ohno-Machado, L. Logistic regression and artificial neural network classification models: A methodology review. *J. Biomed. Inform.* **2002**, *35*, 352–359. [CrossRef]

23. Tao, J.; Klette, R. Integrated pedestrian and direction classification using a random decision forest. In Proceedings of the IEEE International Conference on Computer Vision Workshops, Sydney, Australia, 1–8 December 2013; pp. 230–237.

24. Wang, L. *Support Vector Machines: Theory and Applications*; Springer: Dordrecht, The Netherlands, 2005.

25. Kleinbaum, D.G.; Dietz, K.; Gail, M.; Klein, M.; Klein, M. *Logistic Regression*; Springer: New York, NY, USA, 2002.

26. BigML. Available online: https://bigml.com/ (accessed on 17 December 2019).

27. Microsoft Azure Machine Learning Platform. Available online: https://studio.azureml.net/ (accessed on 17 December 2019).

28. TensorFlow. Available online: https://tensorflow.google.cn/ (accessed on 17 December 2019).

29. Amazon Machine Learning. Available online: https://aws.amazon.com/cn/machine-learning/ (accessed on 17 December 2019).

30. Zhu, X.; Goldberg, A.B. Introduction to semi-supervised learning. *Synth. Lect. Artif. Intell. Mach. Learn.* **2009**, *3*, 1–130. [CrossRef]

31. Ali Humayun, M.; Hameed, I.A.; Muslim Shah, S.; Hassan Khan, S.; Zafar, I.; Bin Ahmed, S.; Shuja, J. Regularized Urdu speech recognition with semi-supervised deep learning. *Appl. Sci.* **2019**, *9*, 1956. [CrossRef]

32. Khan, J.; Lee, Y.K. LeSSA: A unified framework based on lexicons and semi-supervised learning approaches for textual sentiment classification. *Appl. Sci.* **2019**, *9*, 5562. [CrossRef]

33. Wang, X.; Feng, X.; Xu, X.; Diao, G.; Wan, Y.; Wang, L.; Ma, G. Fault plane parameters of Sanhe-Pinggu M8 earthquake in 1679 determined using present-day small earthquakes. *Earthq. Sci.* **2014**, *27*, 607–614. [CrossRef]

34. Fu, C.; Gao, M.; Chen, K. A study on long-period response spectrum of ground motion affected by basin structure of Beijing. *Acta Seismol. Sin.* **2012**, *34*, 374–382.

Article

A Numerical Model for Simulating Ground Motions for the Korean Peninsula

Sang Whan Han and Hyun Woo Jee *

Department of Architectural Engineering, Hanyang University, Seoul 04763, Korea; swhan@hanyang.ac.kr
* Correspondence: hyunwoo71@hanyang.ac.kr; Tel.: +82-2-2220-0566

Received: 7 January 2020; Accepted: 7 February 2020; Published: 13 February 2020

Featured Application: Developing a numerical model for simulating ground motions for low-to-moderate seismic regions such as the Korean Peninsula.

Abstract: Ground motions are used as input for the response history analyses of a structure. However, the number of ground motions recorded at a site located in low-to-moderate seismic regions such as the Korean Peninsula is few. In this case, artificial ground motions need to be used, which can reflect the characteristics of source mechanism, travel path, and site geology. On 15 November, 2017, the Pohang earthquake, with a magnitude of 5.4 and a focal depth of 9 km, occurred near the city of Pohang. This earthquake caused the most significant economic loss among the earthquakes that occurred in the Korean Peninsula. During the Pohang earthquake, valuable ground motions were recorded at stations distributed in the Korean Peninsula. In this study, a ground motion model is proposed based on ground motions recorded during the 2017 Pohang earthquake. The accuracy of the proposed model is verified by comparing measured and simulated ground motions at 111 recording stations.

Keywords: ground motion; earthquake; response history analysis; station; seismicity

---

1. Introduction

The Korean Peninsula is located in stable continental regions about 400 km from the boundaries of four plates: the Philippine Sea, Pacific, North American, and Eurasian Plates [1,2]. The seismic activity in this region is lower than that in regions located near plate boundaries such as Japan, Indonesia, or the Western United States. However, Korean historical documents reported that big earthquakes with a magnitude of 6 or larger occurred in the peninsula from 2 A.D. to 1904 A.D. [3]. The Pohang earthquake, with a magnitude of 5.4 and focal depth of 9 km, occurred near the city of Pohang, located in the Korean South-East province. This earthquake caused the largest casualties and economic loss among the earthquakes that have occurred in Korea.

To protect structures from earthquakes, it is necessary to conduct the seismic performance evaluation of structures with reliable input ground motions, and to retrofit the structures based on the results of the seismic performance evaluation [4–6]. However, it is difficult to collect available ground motions recorded from mid- to large-size earthquakes that have occurred in sites located in low-to-moderate seismic regions. In this case, artificial ground motions can be simulated and used in seismic performance evaluation. Simulated ground motions should retain the characteristics of the local source mechanism, travel path, and geology of a site.

Previous studies [7–12] have developed and improved the ground motion simulation model to generate artificial ground motions by considering key components of seismological characteristics. Ground motions can be simulated using the stochastic point-source model in the frequency domain and the shaping window model in the time domain.

In this study, a ground motion simulation model is proposed. The ground motion parameters are determined based on the ground motions recorded from the mainshock of the 2017 Pohang earthquake.

To verify the accuracy of the proposed model, the key components of measured and simulated ground motions are compared, such as peak ground acceleration (PGA) and 5% damped pseudo spectral acceleration (PSA).

In order to develop the proper ground motion simulation model for the Korean Peninsula, this article is organized into three sections. In Section 2, the collected ground motions in three orthogonal directions (East–West, North–South, and Vertical) are presented, which were recorded from 111 stations in the inland Korean Peninsula during past earthquakes. In Section 3, a stochastic point-source model and shape window model are proposed to simulate the ground motions recorded during past earthquakes that occurred in the Korean Peninsula. In Section 4, the proposed model is verified by comparing the generated artificial ground motions and ground motions recorded during the 2017 Pohang earthquake event.

2. Ground Motions Collected for Developing the Ground Motion Simulation Model

To develop a numerical model for simulating ground motions in the Korean Peninsula, the ground motions recorded at 111 seismic stations during the 2017 Pohang earthquake were collected from the National Earthquake Comprehensive Information System (NECIS) of Korea Meteorological Administration (KMA). Figure 1 shows the distribution of 111 stations out of 157 stations in Korea, which provided ground motion records with a high signal-to-noise ratio (SNR). To exclude ground motions with signal distortion due to noise, we only collected recordings with a signal-to-noise ratio (SNR) greater than 2.0. Each station provided three components (North–South, East–West, and Vertical) of ground motions. Thus, the total number of recorded ground motions is 333 (=3 × 111).

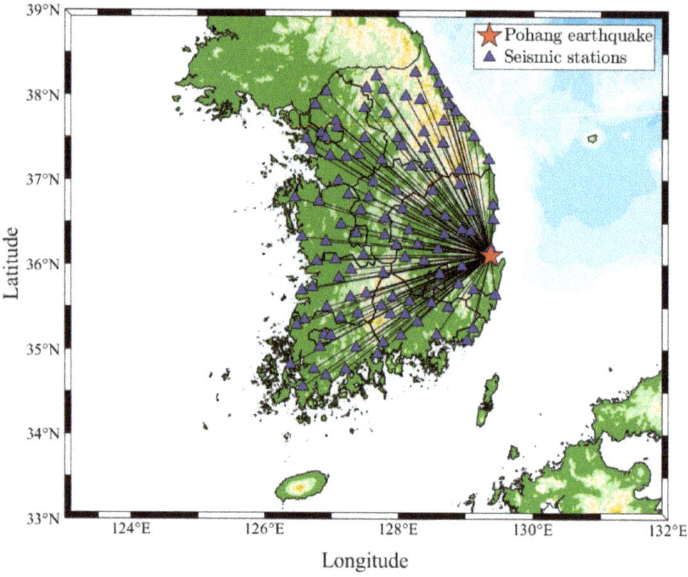

Figure 1. The 2017 Pohang earthquake and seismic recording stations with ray path lines.

All collected ground motions were recorded with a sampling rate of 0.01 s (100 samples per second), and the NECIS provided these records after applying a high cut anti-aliasing filter to remove the noise with a frequency greater than 40 Hz. In this study, low-frequency noise was also removed by using the 0.1 Hz (corresponding period 10 s) low-cut filter, and baseline correction was done using a technique proposed by Papazafeiropoulos and Plevris [13]. Table 1 summarizes the information on the 2017 Pohang earthquake.

Table 1. Information on the 2017 Pohang earthquake (National Earthquake Comprehensive Information System (NECIS)).

Event Name	Local Date-Time	Longitude (East)	Latitude (North)	Focal Depth (km)	Magnitude M_L
the 2017 Pohang earthquake	14:29 15 November, 2017	129.37	36.11	9	5.4

Figure 2 shows the East–West (E–W) direction component of ground acceleration ($\ddot{u}_g$), velocity ($\dot{u}_g$), and displacement (u_g) at station BUS2. Figure 2a–c show raw record data for ground acceleration, velocity, and displacement. Figure 2d–f show the records after applying a low-pass filter, whereas Figure 2g–i show ground motions after applying base line correction. Figure 2j–l show ground motions after applying a low-cut filter, as well as base correction. It can be observed that ground motions are properly adjusted by using both a low-cut filter and base correction.

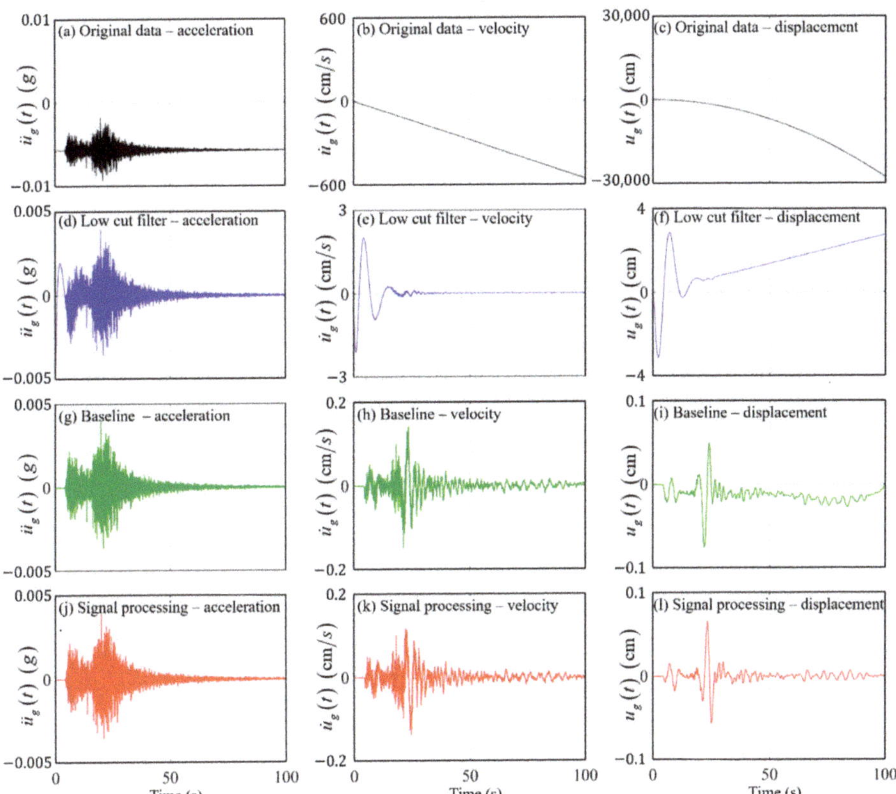

Figure 2. Original and adjusted ground motions at station BUS2: (**a**) Original data-acceleration, (**b**) Original data-velocity, (**c**) Original data-displacement, (**d**) Low cut filter-acceleration, (**e**) Low cut filter-velocity, (**f**) Low cut filter-displacement, (**g**) Baseline-acceleration, (**h**) Baseline-velocity, (**i**) Baseline-displacement, (**j**) Signal processing-acceleration, (**k**) Signal processing-velocity, (**l**) Signal processing-displacement.

The site amplification effect was also removed from the ground motion records because a ground motion model has been developed for hard rock conditions without considering site effects. In general, site effects are considered in numerical models by applying a site amplification factor to the ground motions generated for hard rock site conditions.

In this study, the site amplification effect was removed from the recorded ground motions by using a horizontal-to-vertical spectral ratio (HVSR) technique, which has been widely used for site amplification factor calculation [14–17].

The site amplification factor, $AMP(f)$, can be calculated from Equation (1).

$$AMP(f) = \frac{\sqrt{PSA_{EW}(f) \times PSA_{NS}(f)}}{PSA_V(f)} \tag{1}$$

where $PSA_{EW}(f)$, $PSA_{NS}(f)$, and $PSA_V(f)$ are the 5% damped pseudo spectral acceleration (PSA) from the East–West, North–South, and Vertical components of ground motions, respectively. Figure 3 shows the calculated site amplification factor function estimated from the ground motions recorded at stations BUS2, CEA, and CEJA during the 2017 Pohang earthquake. This function was applied to the Fourier spectrum of a ground motion at each frequency. Figure 4 shows the ground acceleration time histories and Fourier spectra before and after applying Equation (1).

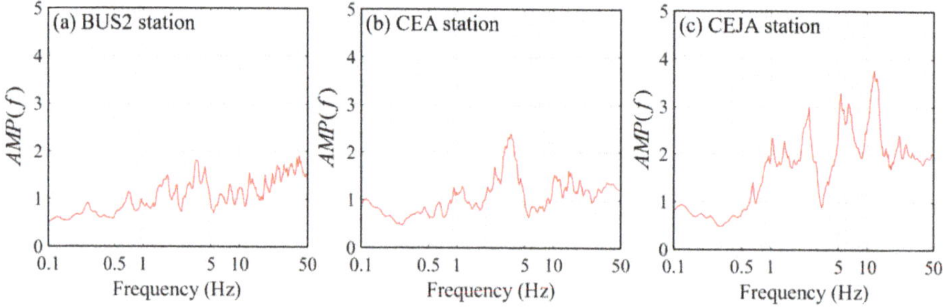

Figure 3. Site amplification factor function from the 2017 Pohang earthquake at sample stations: (**a**) BUS2 station, (**b**) CEA station, (**c**) CEJA station.

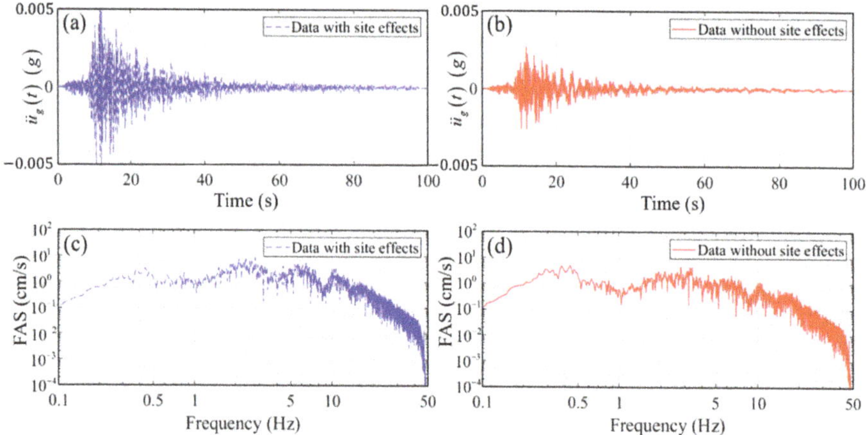

Figure 4. Ground motion recorded at station BUS2 with and without site effects: (**a**) Ground motion acceleration with site effects, (**b**) Ground motion acceleration without site effects, (**c**) Fourier amplitude spectrum with site effects, (**d**) Fourier amplitude spectrum without site effects.

3. Ground Motion Simulation Model

The numerical model to simulate ground motions in the Korean Peninsula is proposed based on the stochastic point-source model in the frequency domain and the shaping window model in the time domain. The point source model is an effective tool for generating ground motions for a site with limited seismological information.

The Fourier amplitude spectrum (FAS) of a ground motion can be determined using the stochastic point-source model [8–12], which was developed based on the omega-square model [18,19]. This model consists of three main seismological characteristics, namely source, path, and site effects [12].

The shaping window model consists of a ground motion envelope and duration time. Saragoni and Hart [7] proposed the envelope of ground motion records, which was adopted in this study. The duration of a ground motion affects the incidence of the collapse of a structure [20].

3.1. Stochastic Point-Source Model Estimation

In the stochastic point-source model, the Fourier amplitude spectrum $[A(f)]$ of a ground motion is calculated using Equation (2).

$$A(f) = Source(M_0, f, f_c) \times Path(R_H, f) \times Site(\kappa_0, f) \tag{2}$$

where $Source(M_0, f, f_c)$ is the earthquake source effect function, $Path(R_H, f)$ is the path effect function, $Site(\kappa_0, f)$ is the site effect function, M_0 is the seismic moment, R_H is the hypocentral distance, f is the frequency, and f_c is the corner frequency. A detailed explanation of individual functions is summarized in Table 2.

The component parameters κ_0, M_0, and f_c to calculate $A(f)$ were estimated from the ground motions recorded at individual stations during the 2017 Pohang earthquake, whereas the values of other component parameters were adopted from previous studies, as listed in Table 2.

To effectively determine the values for κ_0, M_0, and f_c, the $A(f)$ of the recorded ground motions were smoothed using a technique proposed by Konno and Ohmachi [21]. Figure 5 shows the original and smoothed $A(f)$ of ground motions recorded at three sample stations (BUS2, CEA, and CEJA) during the 2017 Pohang earthquake.

Table 2. Information about the functions and component parameters of the stochastic point-source model.

Functions		Parameters
Source effect function	$Source(M_0, f, f_c)$ $= \dfrac{M_0}{1+(f/f_c)^2} \times \dfrac{\langle R_{\theta\phi}\rangle \cdot F \cdot V}{4\pi\rho\beta_s^3} \times (2\pi f)^p \times \dfrac{1}{R_{ref}}$	$\langle R_{\theta\phi}\rangle (= 0.63)$: S-wave averaged radiation pattern coefficient [22] $F(= 2)$: free surface effect [9] $V\left(= 1/\sqrt{2}\right)$: partition coefficient of a vector into the horizontal component [9] $\rho\left(= 2.7 \text{ g/cm}^3\right)$: near source soil density [23] $\beta_S(= 3.36 \text{ km/s})$: near source shear wave velocity [23] p: ground motion type coefficient (0, 1, and 2 for displacement, velocity, and acceleration, respectively) [9] $R_{ref}(= 1 \text{ km})$: reference source-to-site distance for seismic source
Path effect function	$Path(R_H, f)$ $= G(R_H) \times \exp(-\pi f R_H / Q(f)\beta_s)$	$G(R_H) = \begin{cases} R_H^{-1.3} & (R_H \leq 70 \text{ km}) \\ 70^{-1.3} \cdot (R_H/70)^{0.3} & (70 \text{ km} < R_H \leq 100 \text{ km}) \\ 70^{-1.3} \cdot (100/70)^{0.3} \cdot (R_H/100)^{-0.5} & (R_H > 100 \text{ km}) \end{cases}$: geometrical attenuation function [24] $Q(f)\left(= 348 f^{0.48}\right)$: quality factor of the anelastic attenuation function [24]
Site effect function	$Site(\kappa_0, f)$ $= AMP(f) \times \exp(-\pi\kappa_0 f)$	$AMP(f)$: site amplification factor function at each station κ_0: site attenuation coefficient of the site attenuation function [25]

In this study, the site attenuation coefficient, κ_0, was determined first. Since the contribution of κ_0 in Equation (2) is determined without interaction with path effect and site amplification, the contribution of path effect and site amplification were removed from Equation (2). The Fourier amplitude spectrum without path effect and site amplification [$A'(f)$] was calculated using Equation (3).

$$A'(f) = \frac{A(f)}{Path(R_H, f) \times AMP(f)}.$$

(3)

Anderson and Hough [25] estimated the site attenuation coefficient (κ_0) using a residual slope of $A(f)'$ between 10 Hz and 40 Hz, which was also used in this study. Figure 6 shows $A(f)$ and $A(f)'$. The slope (κ_0) of $A(f)'$ is also plotted in Figure 6. The values of κ_0 estimated for individual stations are summarized in Table A1 of Appendix A.

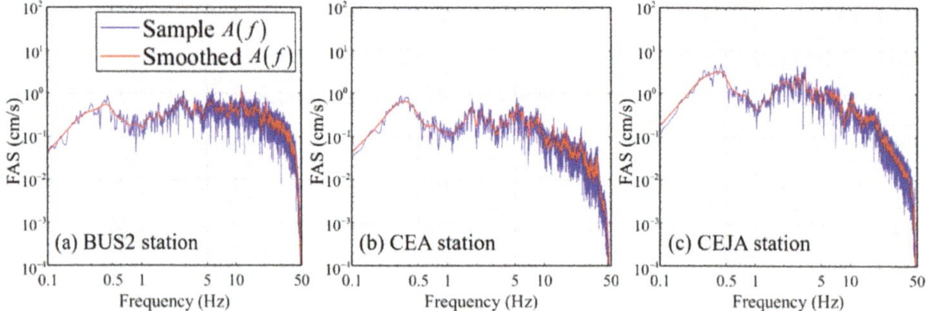

Figure 5. Original and smoothed Fourier amplitude spectrum: (**a**) BUS2 station, (**b**) CEA station, (**c**) CEJA station.

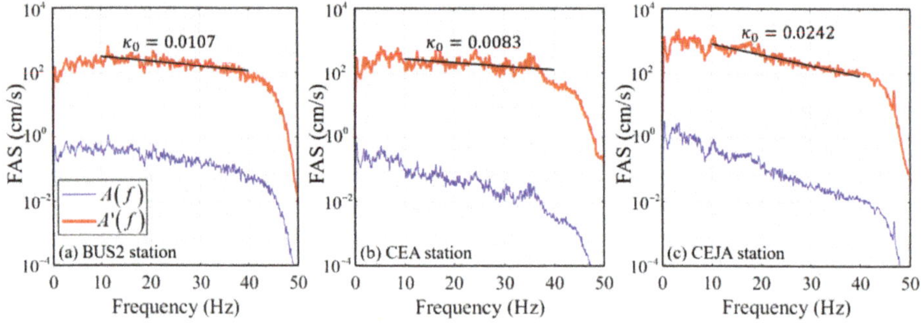

Figure 6. $A(f)$, $A(f)'$, and κ_0 values of three sample stations: (**a**) BUS2 station, (**b**) CEA station, (**c**) CEJA station.

Previous studies [9,26–28] estimated M_o and f_c by using the omega-square model proposed by Brune [18,19] based on the FAS of the recorded ground motions. To determine M_o and f_c without interaction with site attenuation (κ_o), the Fourier amplitude spectrum without the effect of κ_o [$A''(f)$] was calculated using Equation (4). Figure 7 shows $A'(f)$ and $A''(f)$.

$$A''(f) = \frac{A'(f)}{\exp(-\pi\kappa_0 f)}$$

(4)

Corner frequency, f_c, can be estimated using Equations (5)–(8), which were proposed by Andrews [29] and Jo and Baag [27].

$$f_c = \left(\frac{J}{2\pi^3 \Omega_0^2} \right)$$

(5)

$$\Omega_0 = 2\left(\frac{K^3}{J}\right)^{\frac{1}{4}} \tag{6}$$

$$J = \frac{2}{3}(\Omega_0\omega_1)^2 f_1 + 2\int_{f_1}^{f_2}\left|\omega A(\omega)''\right|^2 df + 2\left|\omega_2 A(\omega_2)''\right|^2 f_2 \tag{7}$$

$$K = 2\left|\omega_1 A(\omega_1)''\right|^2 f_1 + 2\int_{f_1}^{f_2}\left|A(\omega)''\right|^2 df + \frac{2}{3}\left|A(\omega_2)''\right|^2 f_2 \tag{8}$$

Seismic moment, M_0, can be also calculated from Equation (9), which was proposed by Joshi et al. [28].

$$M_0 = \frac{4\pi\rho\beta_S{}^3 \times \Omega_0 \times R_{ref}}{\langle R_{\theta\phi}\rangle \times F \times V} \tag{9}$$

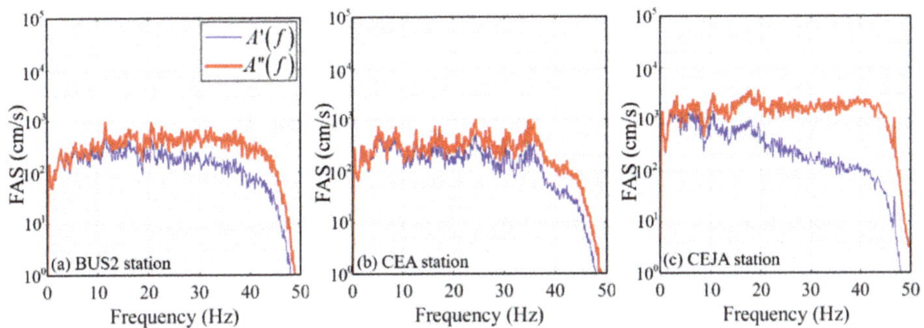

Figure 7. $A'(f)$ versus $A''(f)$ for three sample stations: (**a**) BUS2 station, (**b**) CEA station, (**c**) CEJA station.

Figure 8 shows the values of f_c and M_0 for 111 stations distributed within the Korean Peninsula. Table A2 in Appendix A lists the values of f_c and M_0 estimated for 111 stations. In this study, the median values $(\hat{f}_c, \hat{M}_o)$ of f_c and M_0 were calculated as 0.58 Hz and 8.39×10^{24} dyne-cm, respectively, which were used to calculate $A''(f)$ (Equations (2)–(4)). Figure 9 shows the measured and calculated $A''(f)$ for three sample stations. It can be observed that the calculated $A''(f)$ matches the measured $A''(f)$, by which the accuracy of the proposed procedure is verified.

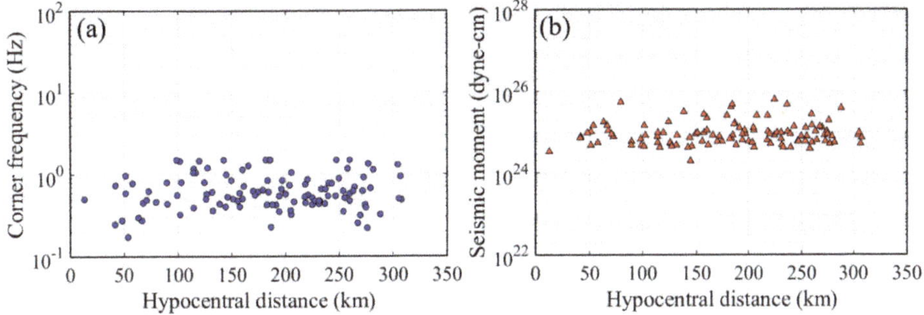

Figure 8. Distributions of the values of f_c and M_0, according to hypocentral distance: (**a**) Measured corner frequencies at 111 stations, (**b**) Measured seismic moments at 111 stations.

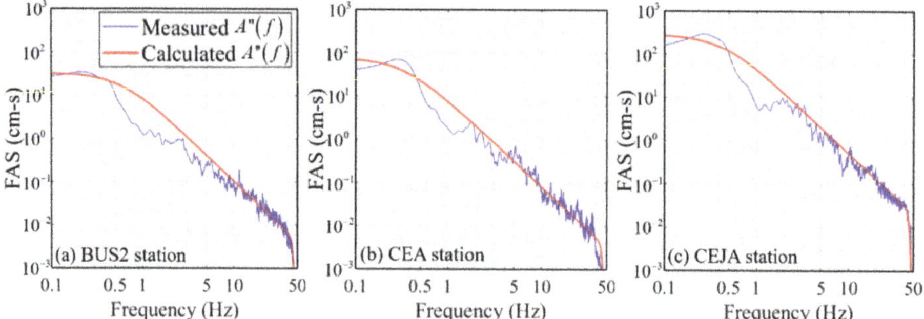

Figure 9. Measured and calculated Fourier amplitude spectrum (FAS) at three sample stations: (**a**) BUS2 station, (**b**) CEA station, (**c**) CEJA station.

3.2. Shaping Window Model Estimation

In order to generate a ground acceleration in the time domain ($\ddot{u}_g(t)$) from $A(f)$, a proper shaping window function should be applied [12]. A shaping window model consists of a ground motion envelope shape and duration time. Boore [12] simulated ground motion recordings using an envelope shape proposed by Saragoni and Hart [7] and a duration time proposed by Atkinson and Boore [11]. However, the shaping window model used by Boore [12] may not properly reflect the envelope shape of the ground motions recorded in the Korean Peninsula. In this study, a shaping window model ($W(t)$) is proposed based on the ground motions recorded during the 2017 Pohang earthquake (Equation (10)).

$$\ln(W(t, T_D)) = c_0 + c_1 \ln\left(\frac{t}{T_D}\right) + c_2 \frac{t}{T_D} \tag{10}$$

where t is the time, and T_D is the duration time. The duration time is calculated from the S-wave arrival time to the time corresponding to the 95% normalized arias intensity energy ($E(t)$). $E(t)$ is estimated using Equation (11).

$$E(t) = \frac{\int_0^t \ddot{u}_g^2(t)dt}{E_T} \tag{11}$$

where E_T is the total energy of $\ddot{u}_g(t)$. Figure 10a–c show the estimated duration time of a ground motion recorded at the BUS2 station, $E(t)$, and estimated shaping window, respectively. Figure 11 shows the estimated shaping window and duration of ground motions recorded at 111 stations during the 2017 Pohang earthquake. The coefficients c_0, c_1, and c_2 in Equation (10) are estimated to match the median envelope shape denoted with a thick solid line in Figure 7a.

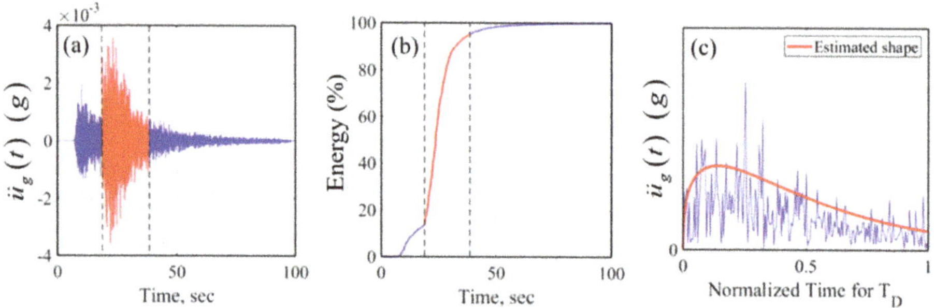

Figure 10. Estimated shaping window and duration at the BUS2 station for the 2017 Pohang earthquake: (**a**) Ground motion acceleration, (**b**) Normalized arias intensity, (**c**) Shaping window.

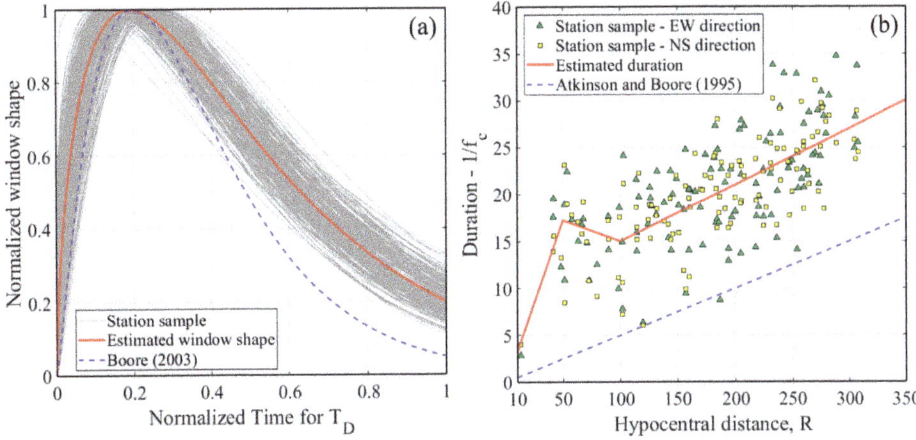

Figure 11. Estimated shaping window and duration for the 2017 Pohang earthquake: (**a**) Shaping window, (**b**) Duration with considering corner frequency.

Regression analyses were conducted, by which the values of c_0, c_1, and c_2 were determined to be 1.6546, 0.6227, and -3.2663, respectively. The envelope shape obtained from the 2017 Pohang earthquake is significantly different from that calculated using the equations proposed by Boor (2003). The envelope shape constructed from the equation proposed by Boor (2003) significantly underestimates the measured envelope shape of ground motions recorded during the 2017 Pohang earthquake.

Figure 11b shows the estimated duration time (T_D) at each station. Previous studies [10,11,30] reported that duration time is mainly affected by corner frequency (f_c) and source-to-site distance (R_H). In this study, the following equation is proposed for T_D, according to R_H and f_c. Figure 11b shows the duration time calculated using Equation (12). This figure also shows that T_D calculated from Equation (12) distinctively differs from that calculated using the equations proposed by Atkinson and Boore [11]. The equation of T_D developed for the northeastern United States could underestimate T_D obtained from the 2017 Pohang earthquake. This observation reveals the importance of why numerical models should be developed for low-to-moderate seismic regions, such as the Korean Peninsula.

$$T_D = \frac{1}{f_c} + \begin{cases} 3.256 & (R_H \leq 10 \text{ km}) \\ -0.247 + 0.350 R_{hypo} & (10 \text{ km} < R_H \leq 50 \text{ km}) \\ -19.522 - 0.045 R_{hypo} & (50 \text{ km} < R_H \leq 100 \text{ km}) \\ 9.005 + 0.060 R_{hypo} & (R_H > 100 \text{ km}) \end{cases} \tag{12}$$

4. Ground Motion Simulation for the 2017 Pohang Earthquake

Ground motions were generated using the proposed numerical model, consisting of the stochastic point-source model and shaping window model. The procedure is briefly summarized as follows:

(1)   White noise is first generated in time domain for a duration time of the ground motion (Equation (12)).

(2)   The noise is then windowed using the shaping window model (Equation (10)).

(3)   The windowed noise is transformed into the frequency domain.

(4)   The $A(f)$ of the noise is normalized by the square-root of the mean square of $A(f)$ in all frequencies.

(5)   The normalized $A(f)$ is then multiplied by the stochastic simulation model (Equation (2)).

(6)   The resulting $A(f)$ is transformed back to the time domain, which is a simulated ground motion.

Figure 12a,b show two horizontal components' ground motions at station BUS2 during the 2017 Pohang earthquake, whereas Figure 12c presents one ground motion simulated using the proposed procedure. Figure 12d presents the geomean of $A(f)$ for recorded ground motions. In this figure, $A(f)$ calculated using Equation (2) is also included, which shows that the calculated $A(f)$ matches that obtained from recorded ground motions. Figure 12e shows the geomean of 5% damped-pseudo spectral acceleration $PSA(T)$ and the median $PSA(T)$ values of 1000 simulated ground motions. The difference in $PSA(T)$ between recorded and simulated ground motions is generally small. Similar observations were made for the ground motions for stations CEA and CEJA.

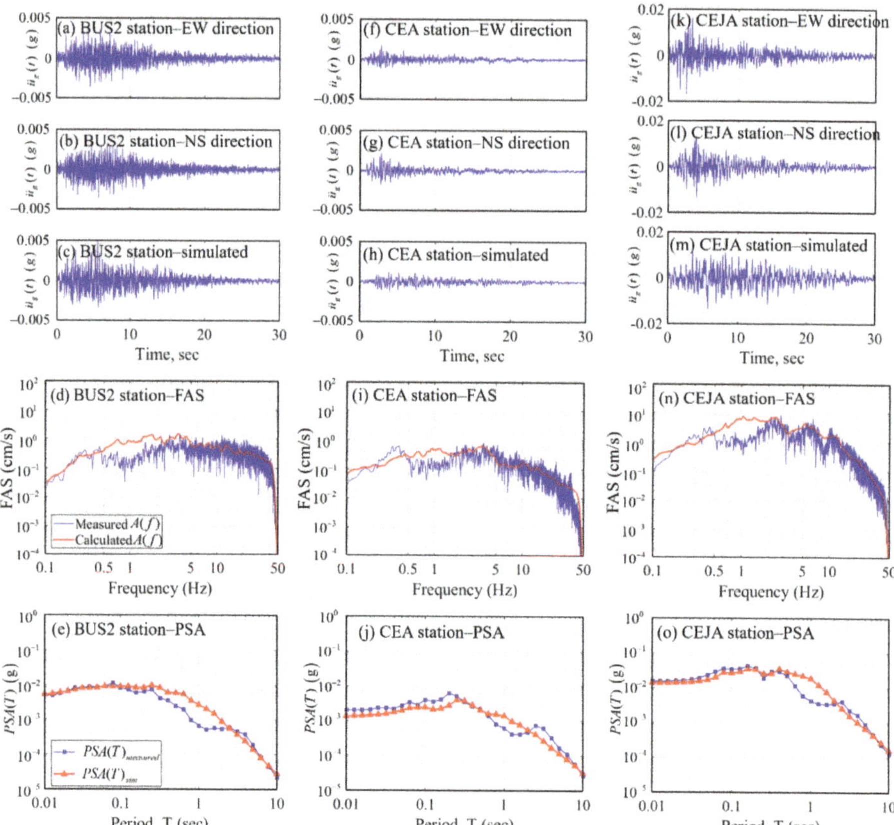

Figure 12. Recorded and simulated ground motion at sample stations: (**a**) BUS2 station-EW direction, (**b**) BUS2 station-NS direction, (**c**) BUS2 station-simulated, (**d**) BUS2 station-FAS, (**e**) BUS2 station-PSA, (**f**) CEA station-EW direction, (**g**) CEA station-NS direction, (**h**) CEA station-simulated, (**i**) CEA station-FAS, (**j**) CEA station-PSA, (**k**) CEJA station-EW direction, (**l**) CEJA station-NS direction, (**m**) CEJA station-simulated, (**n**) CEJA station-FAS, (**o**) CEJA station-PSA.

To verify the accuracy of the proposed numerical model, ground motions were simulated at 111 stations for the 2017 Pohang earthquake and their $PSA(T)$ values were calculated. Residuals induced by the difference in simulated and recorded $PSA(T)$ $(PSA(T)_{sim} PSA(T)_{measured})$ were calculated using Equation (13), with a period (T) range between 0.01 s and 10 s.

$$\text{Residual}(T) = \log[PSA(T)_{sim}] - \log[PSA(T)_{measured}] \tag{13}$$

Residuals were calculated for all ground motions of the 111 stations. Figure 13 shows the residuals according to hypocentral distance. As shown in Figure 13a, the mean value of residuals is near zero, which indicates that the ground motions in the Korean Peninsula can be adequately simulated using the proposed numerical model with the source, path effect, and site amplification functions.

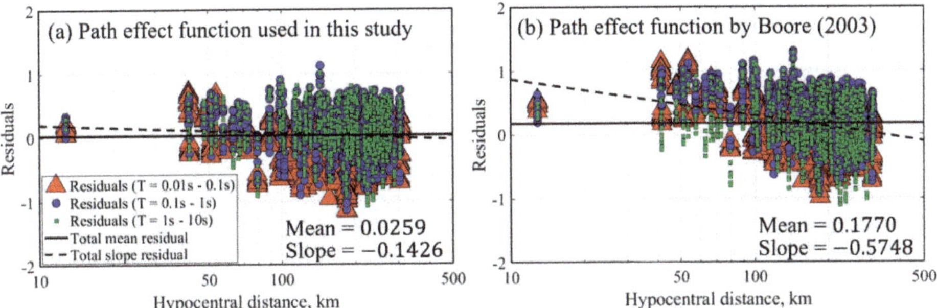

Figure 13. Estimated residuals, mean residual, and slope residual at various periods for the 2017 Pohang earthquake: (**a**) Path effect function used in this study, (**b**) Path effect function by Boore (2003).

If the path effect function proposed by Boor (2003) is used in the numerical model, the mean residual deviates from zero, as shown in Figure 13b. The degree of deviation varies according to the period. In the case of using the source function and site amplification function proposed by others when simulating ground motions with the proposed numerical model, similar observations were made, which are not included in the paper.

5. Conclusions

In this study, a numerical model was proposed to simulate ground motions in the Korean Peninsula. The model consisted of a stochastic point-source model and a shaping window model developed based on ground motions recorded at 111 stations during the 2017 Pohang earthquake. Conclusions obtained from this study are as follows:

(1) Source, path, and site effect functions were developed for the stochastic point-source model, which reflected the seismological characteristics of the Korean Peninsula.
(2) To generate ground motions in the time domain that represents ground motions recorded in the Korean Peninsula, an envelope shape and a duration time function were proposed based on the ground motions recorded at 111 stations.
(3) In order to verify the accuracy of the proposed numerical model, residuals measuring the difference in $PSA(T)$ between recorded and simulated ground motions were calculated for 111 stations. It was observed that ground motions in the Korean Peninsula were simulated accurately using the proposed numerical model, which included proper source, path, and site effect functions.
(4) The results of this study reveal the potential of the proposed numerical model to simulate input ground motions for low-to-moderate seismic regions, such as the Korean Peninsula.

Author Contributions: Methodology, S.W.H.; investigation, H.W.J. All authors have read and agreed to the published version of the manuscript.

Funding: This research was funded by the Korea Agency for Infrastructure Technology Advancement, grant number 20CTAP-C152179-02.

Acknowledgments: This research was supported by grants from the Korea Agency for Infrastructure Technology (20CTAP-C152179-02).

Conflicts of Interest: The authors declare no conflict of interest.

Appendix A

Table A1. κ_0 values for 111 stations.

No.	Station	κ_0	No.	Station	κ_0	No.	Station	κ_0	No.	Station	κ_0
1	ADO2	0.0092	29	GOCB	0.0203	57	JEU2	−0.0035	85	SCHA	0.0161
2	ADOA	0.0284	30	GSG	−0.0112	58	JINA	0.0197	86	SEHB	0.0259
3	BON	0.0182	31	GUM	0.0447	59	JMJ	0.0149	87	SEO2	0.0059
4	BOSB	0.0150	32	GUS	0.0017	60	JNPA	0.0221	88	SES2	0.0132
5	BSA	0.0208	33	GUWB	0.0235	61	JUR	0.0120	89	SHHB	0.0149
6	BURB	0.0254	34	GWJ	0.0289	62	KAWA	0.0227	90	SKC2	0.0102
7	BUS	0.0107	35	GWYB	0.0142	63	KCH2	0.0312	91	SMKB	0.0267
8	BUYB	0.0093	36	HAC	0.0226	64	KMSA	0.0234	92	SUCA	0.0149
9	CEA	0.0083	37	HAD	0.0352	65	KOJ2	0.0241	93	SWO	0.0255
10	CEJA	0.0242	38	HALB	0.0253	66	KWJ2	0.0112	94	TBA2	0.0352
11	CHC2	0.0161	39	HANB	0.0150	67	MAS2	0.0087	95	TEJ2	0.0370
12	CHJ2	0.0111	40	HCNA	0.0245	68	MGY2	0.0177	96	TOHA	0.0183
13	CHO	0.0253	41	HES	0.0167	69	MIYA	0.0287	97	ULJ2	0.0118
14	CHR	0.0254	42	HWCA	0.0192	70	MOP	0.0218	98	USN2	0.0275
15	CHYB	0.0219	43	HWCB	0.0251	71	MUS2	0.0261	99	WJU2	0.0472
16	CIGB	0.0160	44	ICN	0.0131	72	NAJ	0.0186	100	YAPA	0.0208
17	CPR2	0.0077	45	IJA2	0.0197	73	NAWB	0.0122	101	YAY	0.0227
18	CSO	0.0277	46	IJAA	0.0235	74	NOW	−0.0065	102	YAYA	0.0242
19	CWO2	0.0096	47	IKSA	0.0140	75	OKCB	0.0109	103	YCH	0.0179
20	DAG2	0.0197	48	IMSA	0.0235	76	OKEB	0.0269	104	YEG	0.0051
21	DAU	0.0061	49	IMWB	0.0205	77	PHA2	0.0130	105	YEYB	0.0190
22	DGY2	0.0055	50	INCA	0.0125	78	PORA	0.0216	106	YINB	0.0344
23	DUSB	0.0121	51	JAHA	0.0288	79	PTK	0.0113	107	YNCB	0.0214
24	EMSB	0.0279	52	JASA	0.0162	80	PUAA	0.0182	108	YOA	0.0140
25	EURB	0.0130	53	JECB	0.0122	81	PYC	0.0237	109	YOCB	0.0058
26	EUSB	0.0212	54	JEJB	0.0162	82	PYCA	0.0199	110	YODB	0.0222
27	GAPB	0.0288	55	JEO2	0.0021	83	SACA	0.0155	111	YOJB	0.0170
28	GIC	0.0110	56	JES	0.0340	84	SAJ	0.0227		Median	0.0192

Table A2. Calculated corner frequency and seismic moment at all stations used in this study.

No.	$f_{c'}$ Hz	$M_{0,}$ Dyne-cm	No.	$f_{c'}$ Hz	$M_{0,}$ Dyne-cm	No.	$f_{c'}$ Hz	$M_{0,}$ Dyne-cm	No.	$f_{c'}$ Hz	$M_{0,}$ Dyne-cm
1	0.59	1.03×10^{25}	29	0.70	6.37×10^{24}	57	0.44	6.82×10^{24}	85	0.76	2.53×10^{25}
2	0.46	5.42×10^{25}	30	0.22	1.92×10^{25}	58	0.50	2.21×10^{25}	86	0.48	1.46×10^{25}
3	0.84	8.99×10^{24}	31	1.42	6.82×10^{24}	59	0.36	1.27×10^{25}	87	0.25	1.36×10^{25}
4	0.55	5.94×10^{24}	32	0.66	4.21×10^{24}	60	0.64	2.53×10^{25}	88	0.63	8.39×10^{24}
5	1.05	4.21×10^{24}	33	0.49	8.99×10^{24}	61	0.63	1.03×10^{25}	89	0.39	8.99×10^{24}
6	0.43	1.27×10^{25}	34	1.46	5.94×10^{24}	62	0.55	8.99×10^{24}	90	0.35	1.27×10^{25}
7	0.56	4.83×10^{24}	35	0.58	5.94×10^{24}	63	1.37	4.21×10^{24}	91	1.17	3.93×10^{24}
8	0.56	5.17×10^{24}	36	0.44	3.12×10^{25}	64	0.63	5.94×10^{24}	92	0.53	6.37×10^{24}
9	0.32	1.11×10^{25}	37	1.43	8.39×10^{24}	65	0.74	5.54×10^{24}	93	0.89	7.83×10^{24}
10	0.43	4.11×10^{25}	38	0.72	4.83×10^{24}	66	0.78	4.21×10^{24}	94	1.46	9.64×10^{24}
11	0.48	4.83×10^{24}	39	0.49	7.31×10^{24}	67	1.21	4.51×10^{24}	95	1.29	6.82×10^{24}
12	0.50	9.64×10^{24}	40	0.80	1.79×10^{25}	68	0.36	8.39×10^{24}	96	0.59	2.72×10^{25}
13	1.05	4.21×10^{24}	41	0.84	8.99×10^{24}	69	0.80	4.83×10^{24}	97	0.28	1.46×10^{25}
14	1.46	4.51×10^{24}	42	0.58	2.72×10^{25}	70	0.94	5.17×10^{24}	98	0.97	4.83×10^{24}
15	0.32	1.46×10^{25}	43	0.68	7.83×10^{24}	71	1.31	8.39×10^{24}	99	1.37	1.56×10^{25}
16	0.44	5.94×10^{24}	44	0.45	1.03×10^{25}	72	1.23	5.54×10^{24}	100	0.48	9.64×10^{24}
17	0.58	5.17×10^{24}	45	0.66	1.03×10^{25}	73	0.35	6.37×10^{24}	101	0.87	8.99×10^{24}
18	0.74	7.83×10^{24}	46	0.43	6.22×10^{25}	74	0.32	1.27×10^{25}	102	0.44	4.72×10^{25}
19	0.43	1.03×10^{25}	47	0.46	2.53×10^{25}	75	0.40	7.31×10^{24}	103	0.80	8.39×10^{24}
20	0.78	5.54×10^{24}	48	0.66	1.11×10^{25}	76	0.56	7.83×10^{24}	104	1.34	5.17×10^{24}
21	0.63	7.83×10^{24}	49	0.50	5.54×10^{24}	77	0.50	3.42×10^{24}	105	0.30	1.79×10^{25}
22	0.23	1.67×10^{25}	50	0.32	3.84×10^{25}	78	0.63	5.94×10^{24}	106	0.48	9.64×10^{24}
23	0.53	4.51×10^{24}	51	0.92	5.94×10^{24}	79	0.52	8.39×10^{24}	107	0.50	9.64×10^{24}
24	0.84	6.82×10^{24}	52	0.61	5.94×10^{24}	80	0.78	1.36×10^{25}	108	1.14	5.54×10^{24}
25	0.61	4.83×10^{24}	53	0.40	9.64×10^{24}	81	1.27	8.39×10^{24}	109	0.25	7.31×10^{24}
26	0.45	1.11×10^{25}	54	0.17	1.46×10^{25}	82	0.61	4.72×10^{25}	110	0.28	9.64×10^{24}
27	0.52	7.83×10^{24}	55	0.43	4.83×10^{24}	83	0.48	5.94×10^{24}	111	0.39	9.64×10^{24}
28	1.17	5.54×10^{24}	56	1.14	1.11×10^{25}	84	1.05	7.83×10^{24}	Median	0.58	8.39×10^{24}

References

1. Johnston, A.C. Seismic Moment Assessment of Earthquakes in Stable Continental Regions-I. Instrumental Seismicity. *Geophys. J. Int.* **1996**, *124*, 381–414. [CrossRef]
2. Schulte, S.M.; Mooney, W.D. An Updated Global Earthquake Catalogue for Stable Continental Regions: Reassessing the Correlation with Ancient Rifts. *Geophys. J. Int.* **2005**, *161*, 707–721. [CrossRef]
3. Lee, K.; Yang, W.S. Historical Seismicity of Korea. *Bull. Seismol. Soc. Am.* **2006**, *96*, 846–855. [CrossRef]
4. Wen, Y.K.; Collins, K.R.; Han, S.W.; Elwood, K.J. Dual-level designs of buildings under seismic loads. *Struct. Saf.* **1996**, *18*, 195–224. [CrossRef]
5. Lee, L.H.; Lee, H.H.; Han, S.W. Method of selecting design earthquake ground motions for tall buildings. *Struct. Des. Tall. Spec.* **2000**, *9*, 201–213. [CrossRef]
6. Han, S.W.; Seok, S.W. Efficient Procedure for Selecting and Scaling Ground Motions for Response History Analysis. *J. Struct. Eng.* **2014**, *140*, 06013004. [CrossRef]
7. Saragoni, G.; Hart, G. Simulation of Artificial Earthquakes. *Earthq. Eng. Struct. Dyn.* **1974**, *2*, 249–267. [CrossRef]
8. Hanks, T.C.; McGuire, R.K. The Character of High-frequency Strong Ground Motion. *Bull. Seismol. Soc. Am.* **1981**, *71*, 2071–2095.
9. Boore, D.M. Stochastic Simulation of High-frequency Ground Motions Based on Seismological Models of the Radiated Spectra. *Bull. Seismol. Soc. Am.* **1983**, *73*, 1865–1894.
10. Boore, D.M.; Atkinson, G.M. Stochastic Prediction of Ground Motion and Spectral Response Parameters at Hard-rock Sites in Eastern North America. *Bull. Seismol. Soc. Am.* **1987**, *77*, 440–467.
11. Atkinson, G.M.; Boore, D.M. Ground-motion Relations for Eastern North America. *Bull. Seismol. Soc. Am.* **1995**, *85*, 17–30.
12. Boore, D.M. Simulation of Ground Motion Using the Stochastic Method. *Pure. Appl. Geophys.* **2003**, *160*, 635–676. [CrossRef]
13. Papazafeiropoulos, G.; Plevris, V. OpenSeismoMatlab: A New Open-source Software for Strong Ground Motion Data Processing. *Heliyon* **2018**, *4*, e00784. [CrossRef] [PubMed]
14. Nakamura, Y. A Method for Dynamic Characteristics Estimation of Subsurface Using Microtremor on the Ground Surface. *Q. Rep. Railw. Tech. Res. Inst.* **1989**, *30*, 25–33.
15. Atkinson, G.M.; Cassidy, J.F. Integrated Use of Seismograph and Strong-Motion Data to Determine Soil Amplification: Response of the Fraser River Delta to the Duvall and Georgia Strait Earthquakes. *Bull. Seismol. Soc. Am.* **2000**, *90*, 1028–1040. [CrossRef]
16. Zhao, J.X.; Irikura, K.; Zhang, J.; Yoshimitsu, F.; Somerville, P.G.; Asano, A.; Ohno, Y.; Oouchi, T.; Takahashi, T.; Ogawa, H. An Empirical Site-Classification Method for Strong-Motion Stations in Japan Using H/V Response Spectral Ratio. *Bull. Seismol. Soc. Am.* **2006**, *96*, 914–925. [CrossRef]
17. Zandieh, A.; Pezeshk, S. Investigation of Geometrical Spreading and Quality Factor Functions in the New Madrid Seismic Zone. *Bull. Seismol. Soc. Am.* **2010**, *100*, 2185–2195. [CrossRef]
18. Brune, J. Tectonic Stress and the Spectra of Seismic Shear Waves. *J. Geophys. Res.* **1970**, *75*, 4997–5009. [CrossRef]
19. Brune, J. Correction: Tectonic Stress and the Spectra of Seismic Shear Waves. *J. Geophys. Res.* **1971**, *76*, 5002.
20. Chandramohan, R.; Baker, J.W.; Deierlein, G.G. Quantifying the Influence of Ground Motion Duration on Structural Collapse Capacity Using Spectrally Equivalent Records. *Earthq. Spectra* **2016**, *32*, 927–950. [CrossRef]
21. Konno, K.; Ohmachi, T. Ground-Motion Characteristics Estimated from Spectral Ratio between Horizontal and Vertical Components of Microtremor. *Bull. Seismol. Soc. Am.* **1998**, *88*, 228–241.
22. Boore, D.M.; Boatwright, J. Average Body-wave Radiation Coefficients. *Bull. Seismol. Soc. Am.* **1984**, *74*, 1615–1621.
23. Kim, S.K. A Study on the Crustal Structure of the Korean Peninsula. *J. Geol. Soc. Korea* **1995**, *8*, 393–403.
24. Jee, H.W.; Han, S.W. Equations of Path Effects to Simulate Ground Motions in Korean Peninsula Using Point-source Model. In Proceedings of the The 2019 World Congress on Advances in Structural Engineering and Mechanics (ASEM19), Jeju, Korea, 17–21 September 2019.
25. Anderson, J.G.; Hough, S.E. A Model for the Shape of the Fourier Amplitude Spectrum of Acceleration at High Frequencies. *Bull. Seismol. Soc. Am.* **1984**, *74*, 1969–1993.

26. Atkinson, G.M.; Boore, D.M. Evaluation of Models for Earthquake Source Spectra in Eastern North America. *Bull. Seismol. Soc. Am.* **1998**, *88*, 917–934.
27. Jo, N.D.; Baag, C.E. Stochastic Prediction of Strong Ground Motions in Southern Korea. *J. Earthq. Eng. Soc. Korea* **2001**, *5*, 17–26.
28. Joshi, A.; Tomer, M.; Lal, S.; Chopra, S.; Singh, S.; Prajapati, S.; Sharma, M.L.; Sandeep. Estimation of the Source Parameters of the Nepal Earthquake from Strong Motion Data. *Nat. Hazards.* **2016**, *83*, 867–883. [CrossRef]
29. Andrews, D.J. Objective Determination of Source Parameters and Similarity of Earthquakes of Different Size. *Earthq. Source Mech.* **1986**, *37*, 259–267.
30. Boatwright, J.; Choy, G.L. Acceleration source spectra anticipated for large earthquakes in northeastern North America. *Bull. Seismol. Soc. Am.* **1992**, *82*, 660–682.

Article

Study on Seismic Performance of a Mold Transformer through Shaking Table Tests

Ngoc Hieu Dinh, Seung-Jae Lee, Joo-Young Kim and Kyoung-Kyu Choi *

School of Architecture, Soongsil University, 369 Sangdo–ro, Dongjak–gu, Seoul 06978, Korea;
dnhieu@dut.udn.vn (N.H.D.); panderlee@naver.com (S.-J.L.); mzd333@naver.com (J.-Y.K.)
* Correspondence: kkchoi@ssu.ac.kr; Tel.: +82-2-820-0998

Received: 1 December 2019; Accepted: 31 December 2019; Published: 3 January 2020

Abstract: This study presents an experimental seismic investigation of a 1000 kVA cast resin-type hybrid mold transformer through tri-axial shaking table tests. The input acceleration time histories were generated in accordance with the specifications recommended by the International Code Council Evaluation Services Acceptance Criteria ICC-ES AC156 code, with scaling factors in the range of 25–300%. The damage and the dynamic characteristics of the mold transformer were evaluated in terms of the fundamental frequency, damping ratio, acceleration time history responses, dynamic amplification factors, and relative displacement. The shaking table test results showed that the damage of the mold transformer was mainly governed by the severe slippage of the spacers and the loosening of the linked bolts between the bottom beam and the bed beam. In addition, the maximum relative displacement at the top beam in Y and Z-directions exceeded the boundary limit recommended by the Korean National Radio Research Agency. Moreover, the operational test of the specimen was performed based on the IEC 60076-11 Standard before and after the shaking table test series to ensure the operational capacity of the transformer.

Keywords: earthquake/seismic forces; seismic damage; mold transformer; shaking table test; non-structural elements; dynamic characteristics

---

1. Introduction

Non-structural elements that are attached to or supported by structural components play various functions and services in maintaining operation in existing buildings, and to support human activities. According to the complete classification specified in Federal Emergency Management Agency (FEMA) FEMA-74 [1], non-structural elements can be classified into three main categories of architectural components, mechanical and electrical components, and building furnishings and contents. In the building construction, the non-structural elements account for a high percentage of 82–92% of the total economic investment, while structural components account for the remaining 18–8% [2]. Thus, it is obvious that in several vital types of buildings, such as hospitals, high-tech laboratories, power stations, etc., the loss of non-structural elements due to natural disasters could lead to huge replacement costs [3].

During the past few decades, strong earthquake ground motions have caused severe physical, as well as functional, damage to non-structural elements, especially to electrical components, which have led to major operational failures and economic loss of electrical power systems in buildings and special facilities. Depending on the dynamic characteristics, electrical components can be exposed to high-frequency acceleration arising from resonance effects, which result in the loosening of anchor bolts or connecting fasteners, and damage to enclosed plates and frames [4]. For example, the 1994 Northridge earthquake in Los Angeles caused severe damage to crucial non-structural equipment in a major local hospital, such as the emergency power systems, control systems of medical equipment, and water supply piping systems [5]. The 1985 Mexico Earthquake with magnitude 8 and the 2010 Haiti Earthquake with magnitude 7 caused entire failure to unanchored and anchored cabinets in vital facilities [1]. Recently, in South Korea, the 2016 Gyeongju and 2017 Pohang earthquakes caused significant deterioration of non-structural electrical elements within crucial public buildings, such as hospitals, Korea train express (KTX) railway stations, high schools, broadcasting stations, and shopping malls [6]. Thus, nowadays, the investigation of the seismic behavior of non-structural elements is recognized as a key topic in the framework of earthquake risk mitigation.

So far, several research studies have been conducted to evaluate the performance of non-structural elements attached to structural components, subjected to earthquake load. Di Sarno et al. [4] experimentally evaluated the seismic performance of hospital building equipment via unidirectional and bidirectional shaking table testing, considering the presence of internal partitions and cabinet contents. The main purpose of the study was to investigate the correlation between the dynamic response of sample cabinets, and the peak floor accelerations and velocities corresponding to the system limit states, such as overturning and rocking. Based on the test results, fragility curves were also constructed for the components and contents, considering both acceleration and velocity intensity measure. In addition, Petrone et al. [7] performed full-scale shaking table tests in both horizontal directions for the seismic assessment of hollow brick internal partitions. In the study, a steel frame was used to simulate the seismic action at a generic building story and the specimen boundary conditions. The test results showed that the specimen exhibited significant damage at a 0.3% interstorey drift, and extensive damage at a drift close to 1%. In addition, the dynamic characteristics of test specimens were also investigated in terms of damping ratios and natural frequencies, in order to evaluate the influence of the hollow brick partitions on the steel test frame. More recently, Fiorino et al. [8] performed dynamic shaking table tests on prototypes made of indoor partition walls, outdoor facade walls, and suspended continuous ceilings. Based on the test results, the dynamic characteristics in terms of the fundamental frequency and damping ratio, and dynamic amplification were assessed and compared between the non-structural elements with and without anti-seismic solutions. In addition, the seismic response of the tested prototypes was also evaluated in terms of fragility curves.

Nonetheless, the studies on electrical and mechanical components, such as switchboards or transformers commonly used in structural buildings for power systems, are still limited [9–13]. Wang et al. [9] performed a series of quasi-static cyclic loading and shaking table tests to investigate the behavior of a prototype diesel generator equipped with a restrained vibration isolation system. The test results indicated that the failure modes of the restrained isolators were severe fatigue damage of the connection between the vertical restraint rods and top plate, together with the pull-out failure of the vertical restraint rods. In addition, the incorporation of snubbers into the vibration system could provide more restrains, which resulted in effectively preventing the restrained isolators from plastic deformation and severe damage. Hwang et al. [10] performed a seismic fragility analysis of electrical equipment in a typical electric substation in the eastern United States by using actual earthquake damage data. The fragility analysis results revealed the expected performance of electrical equipment in the substation and provided the necessary data for the seismic performance evaluation of an entire electrical substation for reliability analysis. Moreover, Fathali et al. [12] experimentally investigated the seismic performance of an isolation system supporting heavy mechanical equipment. A centrifugal liquid chiller was used as a prototype specimen supported by the American Society of Heating, Refrigerating and Air Conditioning Engineers (ASHRAE)-type isolation system consisting of coil springs and rubber snubbers constraining vertical and horizontal displacement. The test results showed that the isolation system could effectively reduce the response of specimens by energy dissipation and reduce the amplification of the peak acceleration response at the center of the chiller mass with an increase of the peak input acceleration. In addition, the influence of parameters such as the gap size and rubber pad thickness on the seismic performance of the prototype was also analyzed and discussed in detail.

In the present study, dynamic shaking table tests were carried out to investigate the seismic vulnerability of mold transformers, with the aim of expanding the knowledge of the behavior of mechanical and electrical non-structural components. A 1000 kVA hybrid mold transformer was selected as the test specimen, with conventional anchoring details connecting it to a concrete slab. The input acceleration time histories were artificially generated to match the requirements proposed by the International Code Council Evaluation Services Acceptance Criteria ICC-ES AC156 code [14], with different scaling factors. Moreover, random input signals were also used for dynamic system identification, according to FEMA 461 [15]. Based on the test results, the damage stages and dynamic characteristics of the mold transformer during tri-axial acceleration, simulating earthquake load, were evaluated and investigated in terms of the fundamental frequency, damping ratio, acceleration time history responses, and maximum displacement response, as well as dynamic amplification factors.

2. Experimental Program

2.1. Test Specimen

The non-structural electrical component used as a prototype is a hybrid mold transformer that has the advantages of being a high-efficiency transformer and a power-saving function. Such transformers are power saving devices that can help save power in buildings by reducing unnecessary power loss. The hybrid mold transformer used in this study is cast resin-type with the maximum capacity of 1000 kVA and overall dimensions of 2110 mm (height) × 1900 mm (length) × 1030 mm (width). The total mass of the transformer was 3800 kg, according to the data provided by the manufacturer. Figure 1 and Table 1 present a brief description of the major components of the test specimen, which include core, frame system (top beam, bottom beam, and bed beam), high-voltage (HV) coils, low-voltage (LV) coils, and various accessories (lifting lugs, LV and HV terminal, spacer, etc.). Figure 1 shows that the core was made of cold rolled silicon steel and assembled with the frame system via bolt connections; the HV and LV coils were cast in epoxy with a mold under vacuum and were not fixed to the core but were indirectly connected through compressive forces generated from tightened bolts and friction through the spacers.

Table 1. Detailed specifications of the tested mold transformer.

Power Rating (kVA)	Impedance (±10%)	Voltage Regulation (%)	No-Load Current (%)	Standard Efficiency (%)	Length (mm)	Width (mm)	Height (mm)	Operating Weight (kg)
					Dimensions			
1000	4.99	1.3	2.5	99.4	1900	1030	2110	3800

2.2. Test Setup and Measuring Instruments

Figures 2 and 3 show the experimental setup and measuring instruments. The tri-axial tests were carried out using a shaking table, as shown in Figure 4. The main characteristics of the shaking table include: 4.0 m × 4.0 m plan dimensions, six degrees of freedom (SDOF), maximum acceleration of 1.5, 1.5, and 1.0× g in the X, Y, and Z-directions, respectively, maximum pay load of 300 kN, and maximum overturning moment of 1200 kNm. The table is capable of reproducing earthquake input ground motions through a system of eight hydraulic actuators. Table 2 summarizes the specifications of the shake table. The mold transformer test specimen was anchored to a reinforced concrete slab via bed beam systems with eight M16 anchor bolts with diameters of 15.88 mm, according to the manufacturer's installation manual. The concrete slab was connected to the shaking table via M40 anchor bolts with diameters of 40 mm.

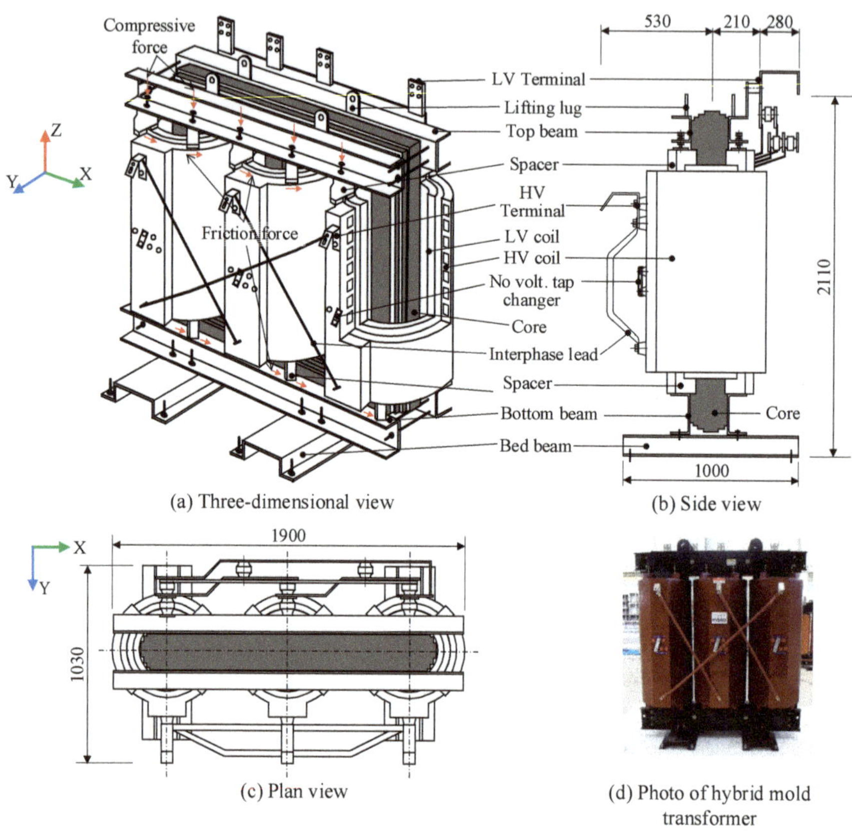

(a) Three-dimensional view

(b) Side view

(c) Plan view

(d) Photo of hybrid mold transformer

Unit: mm

Figure 1. Configurations and details of test specimen. HV, high-voltage; LV, low-voltage: (**a**) Three-dimentional view; (**b**) Side view; (**c**) Plan view; (**d**) Photo of hybrid mold transformer.

Table 2. Detailed specifications of the shaking table used in this test.

Table Size (m × m)	4.0 × 4.0
Type	Fixed
Degrees of freedom	6
Full payload (kN)	300
Overturning moment (kNm)	1200
Acceleration at full payload (× g)	
X-axis	1.5
Y-axis	1.5
Z-axis	1.0
Maximum stroke (mm)	
X-axis	±300
Y-axis	±200
Z-axis	±150
Operational frequency range (Hz)	0.1–60

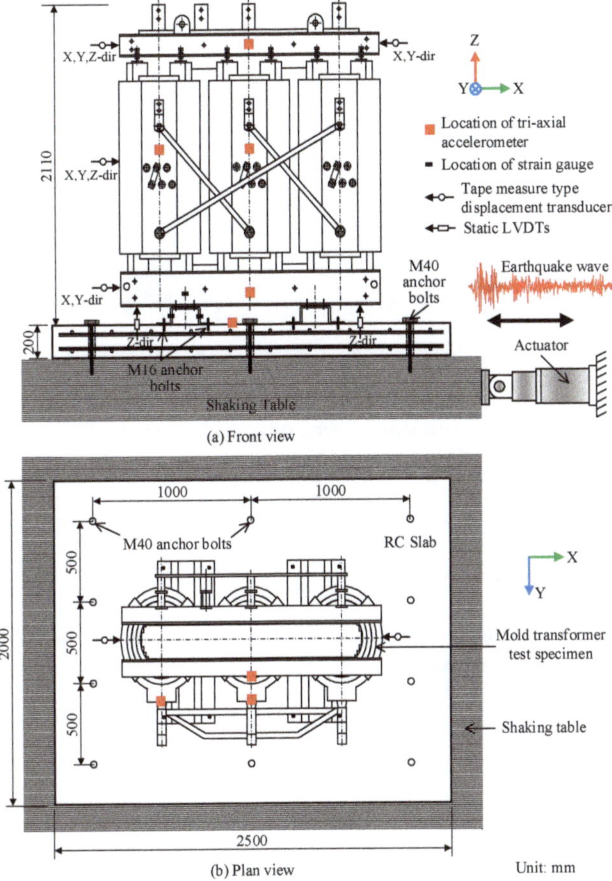

Figure 2. Schematic of test set-up and measurement instrumentation. LVDTs, linear variable displacement transducers: RC = reinforced concrete.

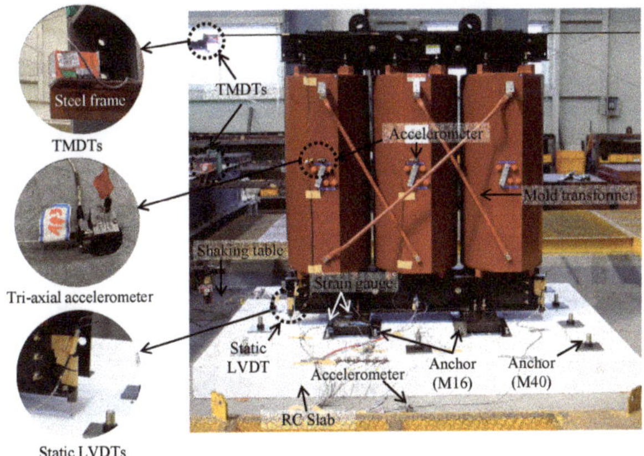

Figure 3. Photograph of test set-up and measurement instrumentation. TMDTs, tape measure type displacement transducers.

Figure 4. Photograph of the shaking table system used in this study.

For measuring instruments, a total of five tri-axial accelerometers were used to record the acceleration response of the test specimen in three orthogonal directions during the tests. Four were mounted on the transformer at the top frame, left side, center zone, and the bed frame, while the fifth was mounted on top of the reinforced concrete (RC) slab (Figures 2 and 3). The accelerometer used in this test has a maximum capacity of ±200× *g*. To measure the mold transformer displacement, tape measure type displacement transducers (TMDTs) and static linear variable displacement transducers (LVDTs) were employed. As shown in Figures 2 and 3, a total of ten TMDTs were fixed on the steel frame out of the shaking table and positioned along the X, Y, and Z-directions at the top left and top right sides of the transformer; and two static LVDTs were positioned along the Z-direction at the bed beam. Furthermore, a total of six steel strain gauges were attached to the bottom beam and bed beam around the locations of linked bolts to monitor the variation of strain during shaking table tests, as shown in Figure 2.

2.3. Input and Testing Protocol

In this study, tri-axial accelerations were generated according to the ICC-ES AC156 code [14]. The input acceleration-time history was artificially generated to match the required response spectrum (RRS) specified by the AC156 code for non-structural components that have fundamental frequencies in the range of 1.3–33.3 Hz. Figure 5 shows that for horizontal RRS, the horizontal spectral acceleration for flexible components, A_{FLX-H}, and horizontal spectral acceleration for rigid components A_{RIG-H}, were determined as:

$$A_{FLX-H} = S_{DS}\left(1 + 2\frac{z}{h}\right) \leq 1.6 S_{DS}, \tag{1}$$

$$A_{RIG-H} = 0.4 S_{DS}\left(1 + 2\frac{z}{h}\right), \tag{2}$$

$$S_{DS} = \frac{2}{3} \cdot F_A \cdot S_S, \tag{3}$$

where S_{DS} is the site-specific ground spectral acceleration factor, defined according to the Korea Building Code [16], (S_{DS} = 0.498 in this study); z and h are the height of the component's attachment point to the structure and the average height of the building roof with respect to the base, respectively; F_A is the site soil coefficient; and S_S is the mapped maximum considered earthquake spectral acceleration at a short period. In the current study, the ratio z/h was equal to 0, with the assumption that the mold transformer was located at the base of the structure.

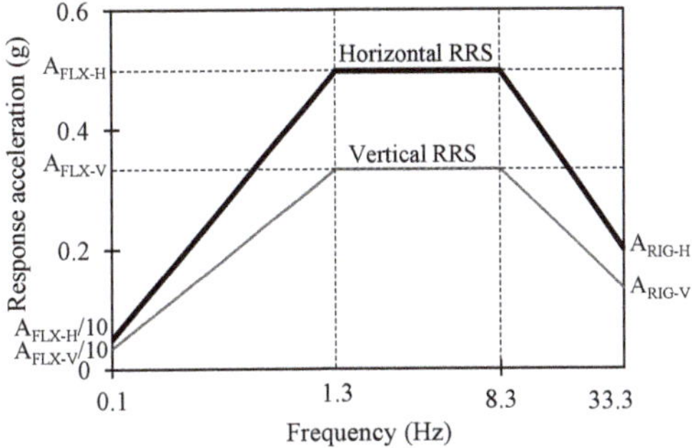

Figure 5. Require response spectrum (AC156 code).

Regarding vertical RRS, the vertical spectral acceleration for flexible components, A_{FLX-V}, and vertical spectral acceleration for rigid components, A_{RIG-V}, were determined as [14]:

$$A_{FLX-V} = 0.67A_{FLX-H},\tag{4}$$

$$A_{RIG-V} = 0.27A_{RIG-H}.\tag{5}$$

Moreover, according to the specifications in ICC-ES AC156 [14] and Eurocode 8 [17], the elastic acceleration spectrum acquired from the selected artificial acceleration time history shall be in a range of 90–130% of RRS, and the matching procedure shall be validated for a range of frequency from 1.3 to 33.3 Hz.

Table 3 summarizes the test program in this study. The test nos. 4–7, 11–14, 18, and 19 are the primary tests. The initial input acceleration time histories in the X, Y, and Z-directions of test no. 7 were artificially generated based on the AC156 code using Equations (1)–(5) and denoted as AC156_100. In the test nos. 4–6, 11–14, 18, and 19, the acceleration magnitudes were scaled from AC156_100 using different scaling factors in the range of 25–300%, corresponding to S_{DS} in the range of 0.12–1.49× g, and denoted as AC156_25 to AC156_300. Figure 6 shows the acceleration time history in the X and Y-directions of AC156_100 used in test no. 7, whereas Figure 7 shows the comparison between the result of the input spectrum of AC156_100 in the X and Y-directions for a damping ratio of 5% and the AC156 target spectrum, as well as its limited boundaries. As shown in the figure, the AC156_100 input spectrum is in a range with a lower limit of 90% RRS and an upper limit of 130% RRS. Table 3 also summarizes the input peak ground accelerations (PGA) of the primary tests corresponding to the scaling factors for each test.

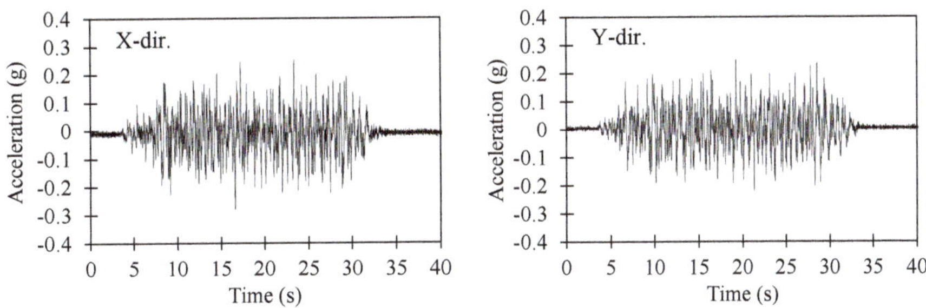

Figure 6. Input acceleration time history in the X and Y-directions for AC156_100.

Table 3. Input and test protocol. PGA, peak ground acceleration.

Test no.	Test Label	Remark	Scaling Factor (%)	Input PGA ($\times g$)		
				X-dir	Y-dir	Z-dir
1–3	Low-amplitude random test	Dynamic identification test. Random input signal, 1–50 Hz				
4	AC156_25		25	0.08	0.07	0.07
5	AC156_50		50	0.15	0.13	0.14
6	AC156_75		75	0.21	0.17	0.18
7	AC156_100		100	0.28	0.25	0.23
8–10	Low-amplitude random test	Dynamic identification test. Random input signal, 1–50 Hz				
11	AC156_125		125	0.33	0.30	0.27
12	AC156_150		150	0.42	0.36	0.31
13	AC156_175		175	0.50	0.43	0.39
14	AC156_200		200	0.57	0.50	0.47
15–17	Low-amplitude random test	Dynamic identification test. Random input signal, 1–50 Hz				
18	AC156_250		250	0.70	0.64	0.58
19	AC156_300		300	0.90	0.79	0.66
20–22	Low-amplitude random test	Dynamic identification test. Random input signal, 1–50 Hz				

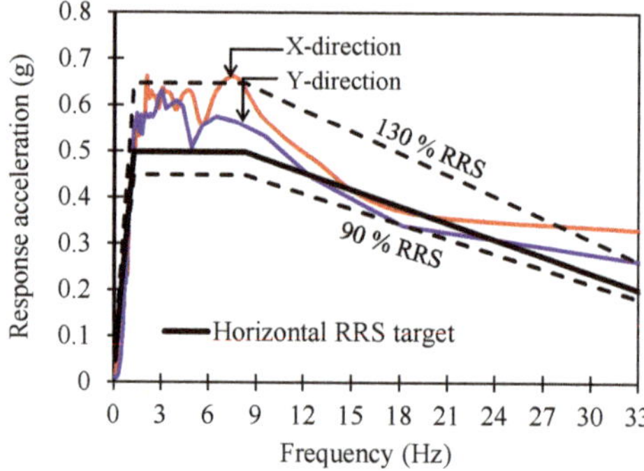

Figure 7. Comparison between the input spectra of AC156_100 and required response spectrum (RRS) target.

Along with the primary tests, the intermediate tests (test nos. 1–3, 8–10, 15–17, and 20–22) were carried out for dynamic identification of the test specimen. Such tests were performed in the X, Y, and Z-directions by applying low-amplitude random input signals with the frequency domain in the range of 1–50 Hz, according to FEMA 461 [15]. To be more specific, test nos. 1–3 were carried out before the AC156_25 (test no. 4) in the X, Y, and Z-direction, respectively; test nos. 8–10 were carried out after the AC156_100 (test no. 7) in the X, Y, and Z-direction, respectively; test nos. 15–17 were carried out after the AC156_200 (test no. 14) in the X, Y, and Z-direction, respectively; and test nos. 20–22 were carried out after the AC156_300 (test no. 19) in the X, Y, and Z-direction, respectively. Note that each dynamic identification test had the same peak acceleration amplitude of approximately $\pm 0.2 \times g$, but had different acceleration time history.

3. Test Results and Discussion

3.1. Dynamic Identification

The dynamic properties, including the natural frequencies, f, and damping ratios, ξ, of the test specimen were investigated in this study. The acceleration responses obtained from dynamic identification tests were analyzed to identify the dynamic properties of the test specimen in both the horizontal and vertical directions. The fundamental frequencies were evaluated based on the transfer function method in the frequency domain. The transfer function amplitude was determined as the ratio between the Fourier transformation of the input signals, and the response output signals collected from accelerometer data installed on the several points of the mold transformer [7,8,11,18]. The sampling frequency of the accelerometer in this study was equal to 512 Hz, and the size of each data block (window) was set to 5120, corresponding to 10 s. The transfer function amplitude has local peaks at the natural frequency of the system [18].

Figure 8 illustrates the dynamic identification results and transfer function curves evaluated from the data recorded at the top beam. To eliminate the noise from the experimental results and obtain the fitting curves, the estimation algorithm was used to obtain the continuous-time transfer function model using time-domain data from the input and output signals [19]. From the Fourier transformation results, the data gathering showed an inefficient trend in the high-frequency domain, due to the fluctuation. Therefore, to get an effective transfer function model, the frequency domain was defined as below 30 Hz to filter the input and output data.

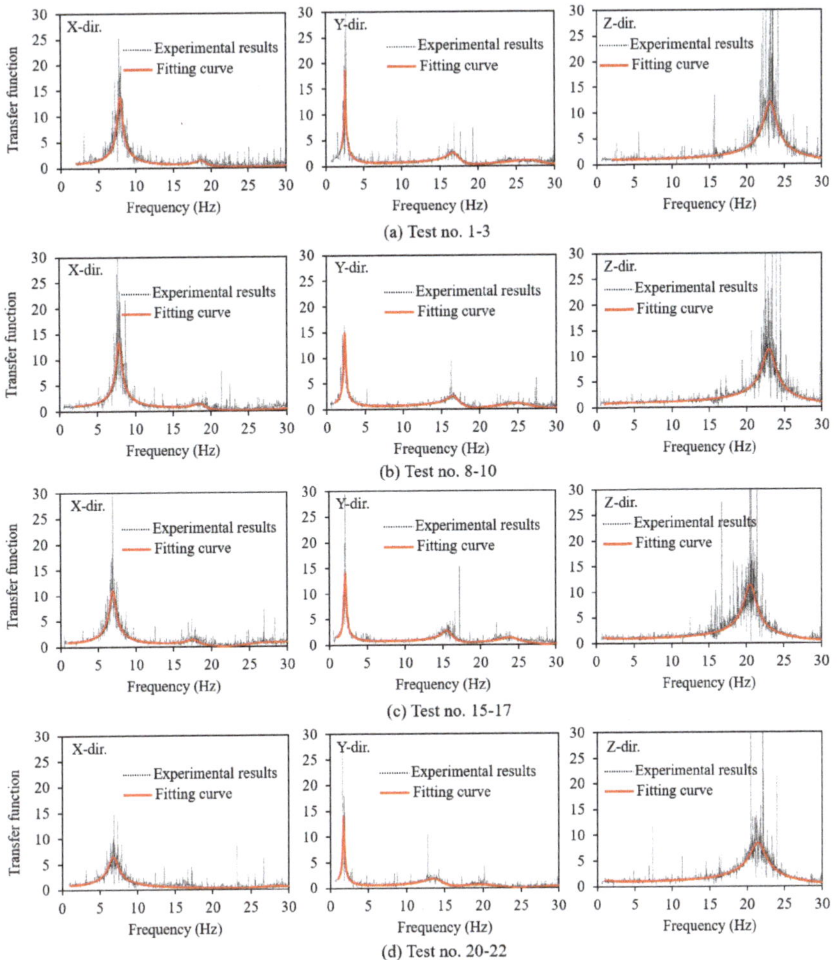

Figure 8. Transfer function amplitude versus frequency curves at top beam: (**a**) Test no.1–3; (**b**) Test no.8–10; (**c**) Test no.15–17; (**d**) Test no.20–22.

Figure 9 shows the results of the fundamental frequency as a function of the scaling factor at different locations of the test specimen: the top beam, the left coil, and the center coil. In general, the initial fundamental frequencies of the mold transformer were almost the same regardless of the location, and those in the X, Y, and Z-directions were 7.87, 2.52, and 23.12 Hz, respectively. In the Y-direction, the specimen showed a low natural frequency, which indicated low stiffness in this direction. The fundamental frequency in the Z-direction was much larger than those in the X and Y-directions. This was attributed to the contribution of axial stiffness of all anchors in the vertical direction, leading to the high stiffness of the test specimen. Similar test results and trends were observed in the previous study by Wang et al. [9], despite the discrepancy in the electrical testing prototype. In Figure 9, since the damage grew as the input acceleration amplitude increased, the fundamental frequency shows a decreasing trend; however, the level of frequency decline was not significant.

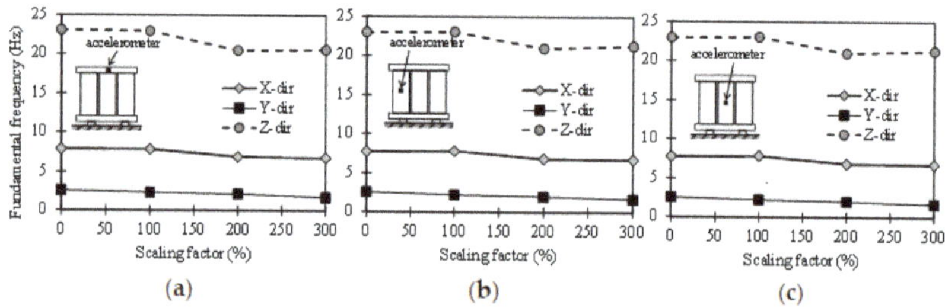

Figure 9. Variation of fundamental frequency of test specimen evaluated based on acceleration data at different locations. (**a**) At top beam; (**b**) At left coil; (**c**) At center coil.

Furthermore, after determining the transfer function curves in the frequency domain, the damping ratios were then calculated at a given resonant of frequency using the half-power bandwidth method [20–22], as follows:

$$\xi = \frac{f_2 - f_1}{2f_o} \times 100\%,\tag{6}$$

where f_o (Hz) is the frequency at the peak transfer function amplitude, and f_1 and f_2 (Hz) are the frequencies associated with the amplitude that is $\left(1/\sqrt{2}\right)$ times lower and higher than the peak transfer function amplitude, respectively. Figure 10 presents the variation of damping ratio measured at several points on the test specimen according to the increase of scaling factor. Overall, the initial damping ratio of the test specimen was in the range of 2–4%. The damping ratio increased to 4–10%, along with the increase of input ground motion amplitude due to the damage accumulated in the transformer as the input acceleration amplitude increased. The analogous results were observed in the previous study by Fathali [12,13] on non-structural electrical equipment.

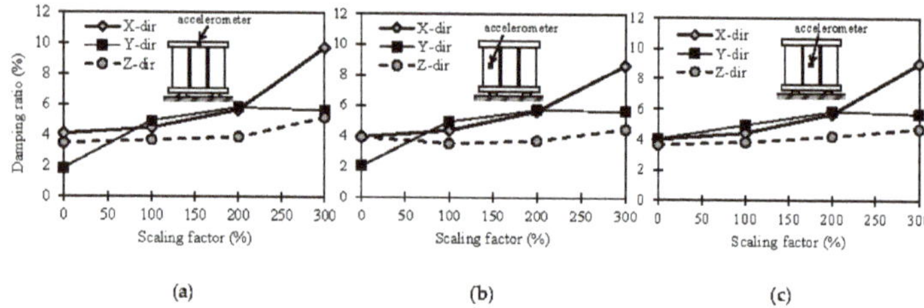

Figure 10. Variation of damping ratio of test specimen evaluated based on acceleration data at different locations. (**a**) At top beam; (**b**) At left coil; (**c**) At center coil.

3.2. Damage Observation

Figure 11 presents the typical damage of test specimen observed during a series of shaking table tests. Overall, three weak points are shown, as evident in Figure 11a: The failure of the spacers, the slippage of coils, and the loosening of linked bolts between the bottom beam and bed beam. The first damage was observed after AC156_50 (Test no. 5), with respect to the PGA of 0.15, 0.14, and 0.16× g in the X, Y, and Z-directions, respectively. The local damage could be observed in the spacers located on the right and left coils of the transformer in terms of vertical and horizontal crack, as shown in Figure 11b.

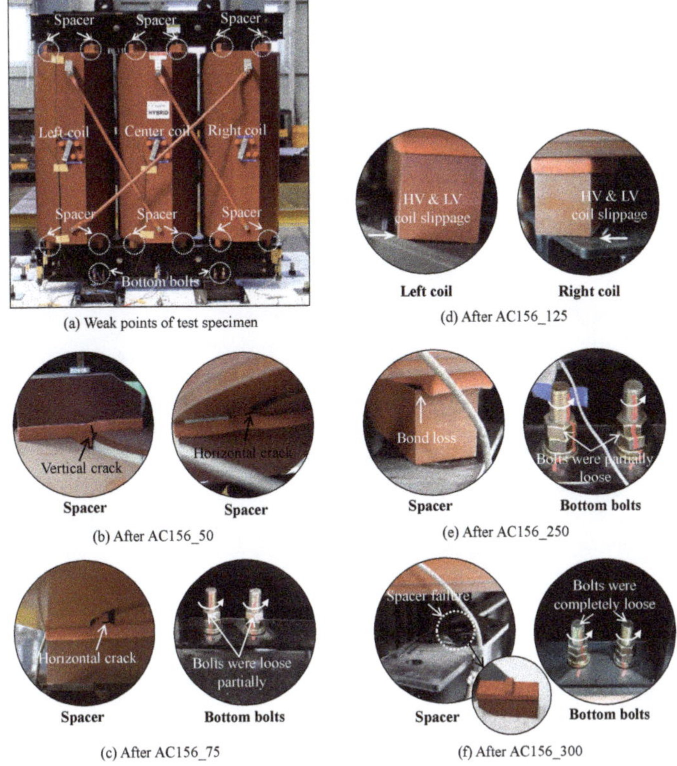

Figure 11. Damage observation after different shaking table tests: (**a**) Weak points of test specimen; (**b**) After AC156_50; (**c**) After AC156_75; (**d**) After AC156_125; (**e**) After AC156_250; (**f**) After AC156_300.

Figure 11c–f demonstrate the damage of the test specimen observed after AC156_75, AC156_125, AC156_250, and AC156_300, respectively. To be more specific, after AC156_75 (Test no. 6), with respect to the PGA of 0.21, 0.17, and 0.18× g in the X, Y and Z-directions, respectively, the horizontal crack continues to develop in the spacers located on the left coil at the bottom of the transformer (Figure 11c). Simultaneously, the bolts connecting the bottom beam and bed beam partially loosen. After AC156_125 (Test no. 11), with respect to the PGA of 0.33, 0.31, and 0.27× g in the X, Y and Z-directions, respectively, the HV and LV coils have slipped away from the original positions (Figure 11d). This is due to the fact that the inertia force caused by the high acceleration level exceeds the friction force between the spaces and bottom beams, leading to the slippage of the HV and LV coils. The slippage of the HV and LV coils continued to grow during test AC156_250 and caused bond loss between the upper part and lower part of the spacers (Figure 11e). After the final test (AC156_300), with respect to the PGA of

0.90, 0.79, and 0.66× g in the X, Y, and Z-directions, respectively, the bond between the upper part and lower part of the spacers located at the bottom of the transformer was completely lost, leading to the failure of the spacers. Simultaneously, the bolts connecting the bottom beam and bed beam became completely loose, as shown in Figure 11f.

In general, at the final test, the specimen neither overturned nor collapsed; however, the spacers showed severe slippage, which to avoid magnetic stress and dangerous geometrical dissymmetry, should keep constant distances between the core and the coils, and between the HV coils and LV coils [23]. Such failure of the spacers was mainly concentrated at the bottom region of the test specimen, along with the loosening of the linked bolts between the bottom beam and bed beam.

3.3. Acceleration Response and Dynamic Amplification of Test Specimen

The tri-axial acceleration response histories measured at different locations (top beam, bottom beam, left coil, and center coil) of the mold transformer were used to analyze the test results. Table 4 summarizes the peak acceleration responses of the test specimen in the X, Y, and Z-directions. Figures 12 and 13 present the acceleration response time histories of the test specimen subjected to 75% tri-axial AC156 input ground motion (AC156_75 test) and 300% tri-axial AC156 input ground motion (AC156_300 test), respectively. Figure 12 shows that the acceleration responses at the center coil and the left coil of the test specimen were almost the same. In the Y-direction, where the specimen showed low stiffness, the acceleration response showed a big difference, compared to those in the X and Z-directions, the top beam vibrated severely with 0.7× g being the largest value of response acceleration; in general, the bottom beam showed less vibration than the other locations; however, transient peak accelerations were recorded with a maximum value of 0.51× g. Figure 13 shows that the specimen also revealed almost the same acceleration responses at the center coil and the left coil. In the Y-direction, the trend was similar to the test of AC156_75, the top beam vibrated more severely than the other locations with a peak response acceleration of 3.30× g; overall, at the bottom beam the level of vibration was not different from those of the left coil and center coil; however, transient peak accelerations were recorded with a maximum value of 2.31× g. This can be attributed to the high amplitude of input ground motion, which caused the damage accumulation in the bottom beam and the shift of natural frequency, as well as damping ratio, as presented in the aforementioned section.

Table 4. Peak response acceleration of the test specimen at different locations.

Test no.	Test Label	Peak Response Acceleration (× g)											
		Top Beam			Center Coil			Bottom Beam			Left Coil		
		X	Y	Z	X	Y	Z	X	Y	Z	X	Y	Z
4	AC156_25	0.17	0.19	0.10	0.20	0.15	0.13	0.08	0.11	0.09	0.18	0.12	0.13
5	AC156_50	0.34	0.36	0.20	0.33	0.24	0.24	0.16	0.17	0.21	0.36	0.24	0.22
6	AC156_75	0.45	0.70	0.42	0.42	0.36	0.36	0.23	0.51	0.33	0.42	0.40	0.34
7	AC156_100	0.54	0.94	0.54	0.48	0.49	0.43	0.54	0.45	1.04	0.52	0.53	0.46
11	AC156_125	0.70	1.09	1.20	0.59	0.52	0.61	0.96	1.10	1.00	0.51	0.69	0.60
12	AC156_150	0.75	0.98	0.90	0.62	0.59	0.76	0.71	1.03	1.25	0.60	0.67	0.74
13	AC156_175	0.86	1.05	1.35	0.68	0.69	0.70	0.84	1.27	1.68	0.77	0.69	0.83
14	AC156_200	1.05	1.33	0.72	0.84	0.85	0.72	0.82	1.50	1.44	0.95	0.89	1.02
18	AC156_250	1.48	1.84	0.86	0.99	1.10	0.82	1.49	1.90	1.87	1.23	1.47	1.18
19	AC156_300	1.62	3.30	1.11	1.21	1.10	1.02	1.57	2.31	1.11	1.28	1.41	1.58

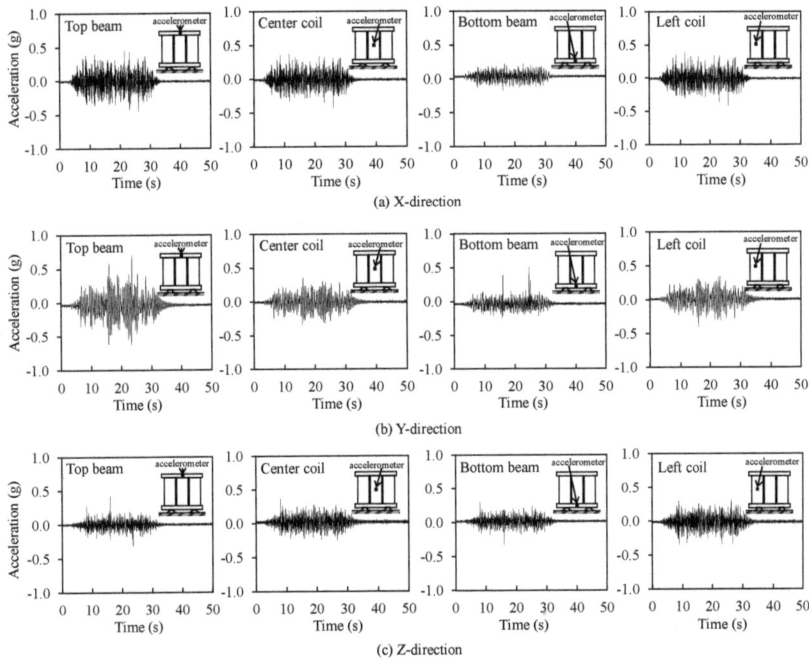

Figure 12. Acceleration response-time histories of test specimen of AC156_75.

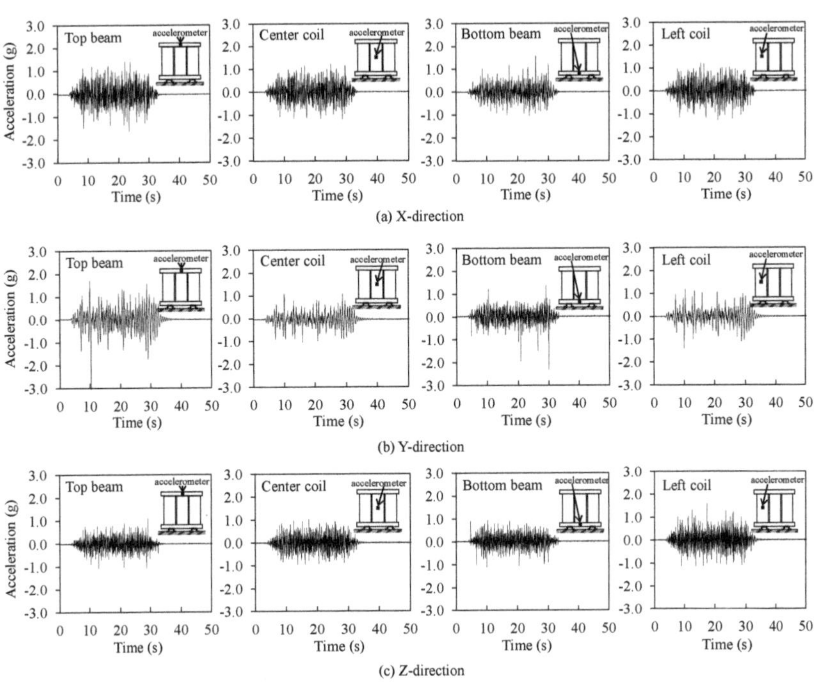

Figure 13. Acceleration response-time histories of test specimen of AC156_300.

Figure 14 presents the results of the peak response acceleration of the test specimen with respect to peak input accelerations at different locations. Overall, it can be observed that the specimen showed almost similar peak response acceleration, regardless of location in the X-direction. However, in the Y and Z-directions, it showed big differences in response acceleration at different locations. In the Y-direction, after a PGA of around 0.2× g, the peak response acceleration at the top beam and bottom beam were higher than those of the center coil and left coil and reached values of 3.30 and 2.31× g at AC156_300, respectively. Meanwhile, the peak acceleration responses at the center and left coils were 1.10 and 1.47× g, respectively. This is because the center and left coils are partly isolated due to the epoxy and vibrated separately from the steel frame core of the transformer. Moreover, Figure 14a also presents the damage stages of the test specimen: the local damage of spacers was observed at a PGA of around 0.15× g, the bolts connecting the bottom beam and bed beam were partly loose at a PGA of around 0.20× g, the slippages of HV and LV coils were observed at PGA of around 0.30× g, and the bolts connecting the bottom beam and bed beam were completely loose at a PGA of around 0.60× g.

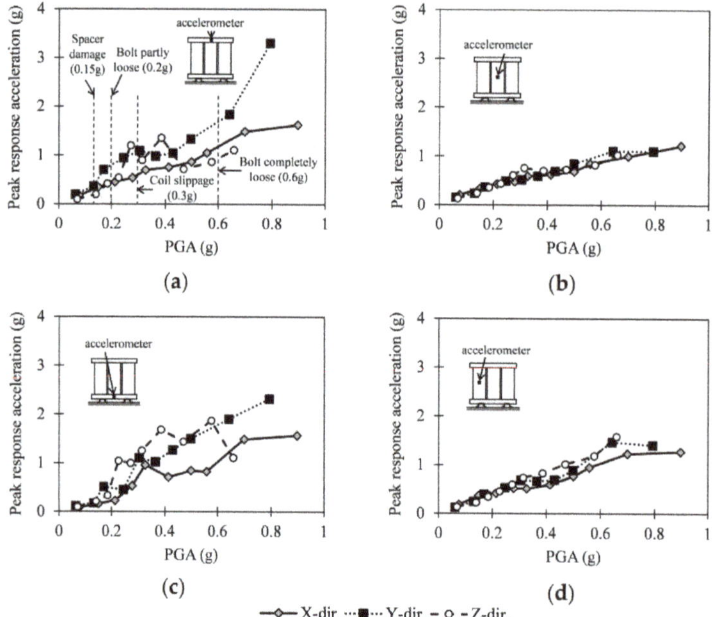

Figure 14. Peak acceleration response versus PGA at different locations of accelerometers. (**a**) At top beam; (**b**) At center coil; (**c**) At bottom beam; (**d**) At left coil.

In addition, the dynamic amplification was evaluated, which is the key parameter in seismic engineering of non-structural components. In this study, the dynamic amplification of the mold transformer can be evaluated by means of the acceleration amplification factor, a_P, which was defined as the ratio between the peak response acceleration of the test specimen (PRA), and the peak floor acceleration (PFA) [4,8]:

$$a_P = \frac{\text{PRA}}{\text{PFA}}. \tag{7}$$

In Equation (7), the values of PRA were obtained from the accelerometers mounted on the specimens and the values of PFA were obtained from the accelerometers mounted on top of the RC slab, as shown in Figures 2 and 3. Figure 15 illustrates the values of the amplification factor according to the peak ground acceleration calculated at the top beam, center coil, bottom beam, and left coil of the test specimen with respect to the X, Y, and Z-directions. According to the recommendations of

FEMA E-74 [1] and American Society of Civil Engineers (ASCE) ASCE 7-16 [24], the design component amplification factor varies from 1.0 (for rigid components) to 2.5 (for flexible components), which is also presented in Figure 15 for comparison. Overall, the amplification factors in the X and Y-directions were in the range of those for non-structural elements specified in FEMA E-74 [1]. Meanwhile, in the Z-direction, the acceleration amplification factors were smaller than the lower limit specified in FEMA E-74 [1]. From the study by Fathali et al. [12,13] on the seismic performance of electrical components, the amplification factors in the horizontal and vertical directions were almost the same, which showed a discrepant trend to the present results. This could be attributed to the discrepancy of anchoring details connecting the transformer to the concrete slab, resulting in the different response acceleration of the test specimen in horizontal and vertical directions.

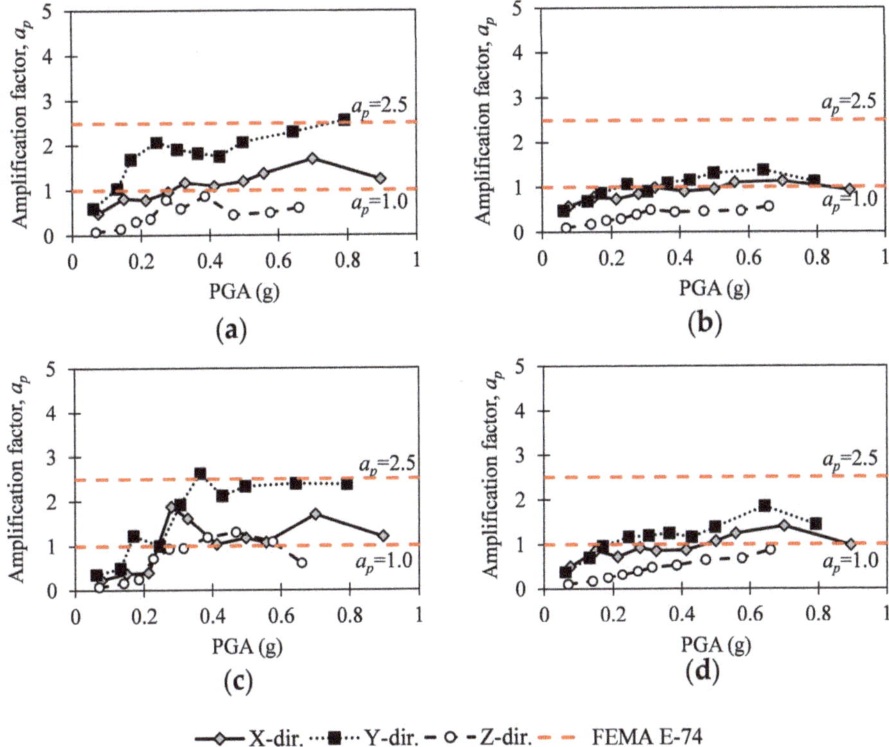

Figure 15. Amplification factor of test specimen at different locations of accelerometers. (**a**) At top beam; (**b**) At center coil; (**c**) At bottom beam; (**d**) At left coil.

3.4. Displacement Response of Test Specimen

The data recorded from the TMDTs and static LVDTs were calibrated to determine the relative displacement response of the test specimen. The tri-axial relative displacement response at a specified location of the transformer could be derived from a system of quadratic equations:

$$\begin{cases} (x - x_0)^2 + y^2 + z^2 = r_x^2 \\ x^2 + (y - y_0)^2 + z^2 = r_y^2 \\ x^2 + y^2 + (z - z_0)^2 = r_z^2 \end{cases}, \tag{8}$$

where x, y, and z are the calculated relative displacements of a specified location of the transformer in X, Y, and Z-directions, respectively; x_o, y_o, and z_o are the absolute distances between the fixed locations of TMDTs to the measured locations of the transformer in X, Y and Z-directions, respectively; and r_x, r_y, and r_z are the absolute displacement values recorded from the TMDTs in X, Y and Z-directions, respectively. Figure 16 expresses the relative displacement response-time histories at the left coil of the test specimen subjected to 75% tri-axial AC156 input ground motion (AC156_75 test) and 300% tri-axial AC156 input ground motion (AC156_300 test). Figure 17 expresses the tri-axial maximum relative displacement response of the test specimen evaluated at the top beam, left coil, and bottom beam of the test specimen with respect to the PGA. In general, the maximum relative displacement of the test specimen increased along with the increase of PGA. The maximum relative displacement in the Z-direction was much smaller than those in the X and Y-directions during the shaking table tests.

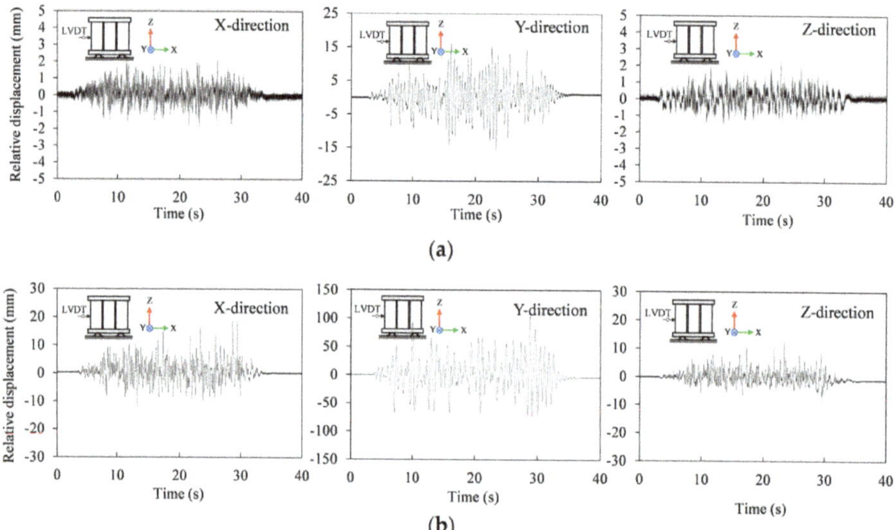

Figure 16. Relative displacement response-time histories of test specimen at the left coil. (**a**) at AC156_75; (**b**) at AC156_300.

Moreover, according to the provisions and recommendations of the Korean National Radio Research Agency [25], the maximum displacement at the top of the equipment should not exceed 75 mm to ensure the safety and functional operation of non-structural components, as well as adjacent components. Figure 17 also shows the limit condition of 75 mm in comparison with the test results. The figure shows that the maximum relative displacements in the Z-direction do not exceed the boundary limit of 75 mm at the end of the tests. Meanwhile, the maximum relative displacements in the X and Y-directions exceed the boundary limit of 75 mm around the PGA of 0.70 and 0.5× g, respectively.

3.5. Strain Profiles

Figure 18 presents the maximum strain profiles recorded at the locations around the linked bolts of the bottom beam and the bed beam during the shaking table test series, with respect to the scaling factor. Overall, the strain of the bottom beam and bed beam increased along with the increasing input acceleration amplitude but did not exceed the yield strain. At the bottom beam (Figure 18a), the maximum measured strain was 0.00129 mm/mm at AC156_300. At the bed beam (Figure 18b), the maximum measured strain was 0.00198 mm/mm at AC156_300, which nearly reached the yield strain of 0.002 mm/mm.

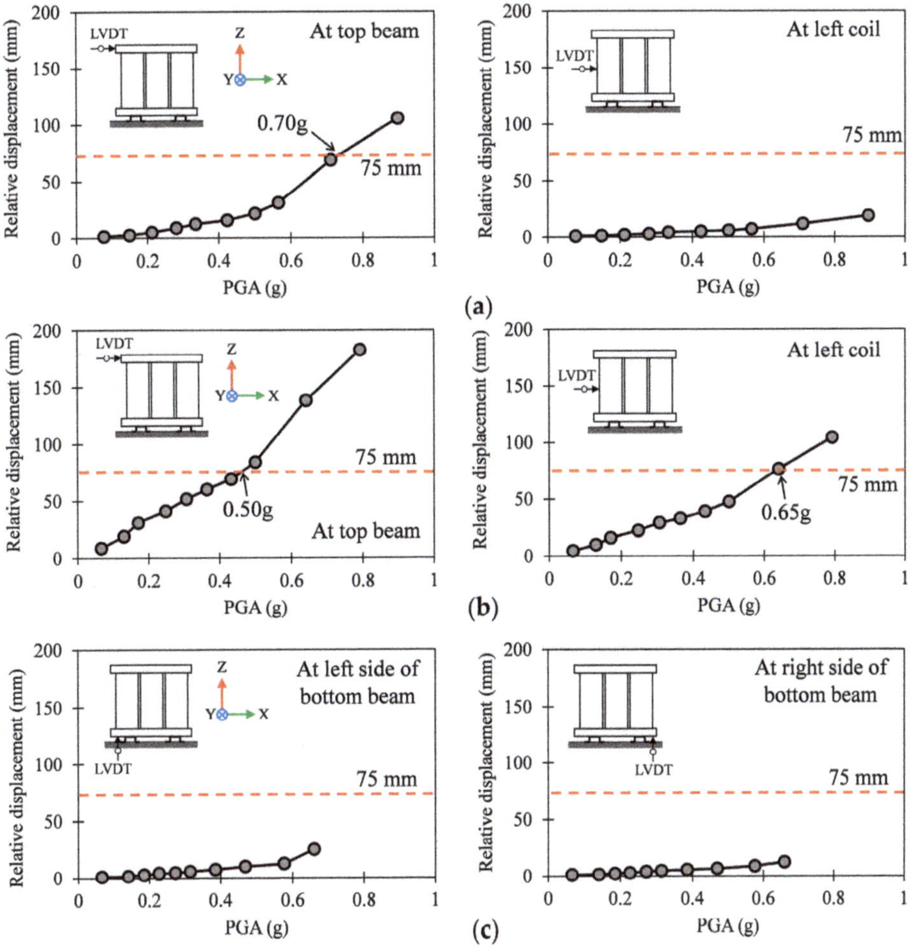

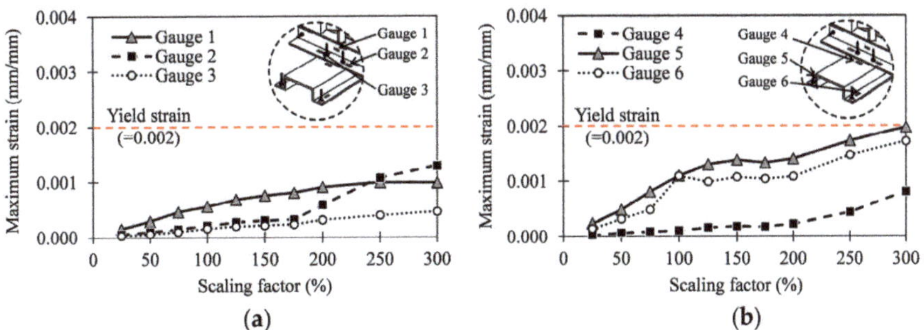

Figure 17. Maximum relative displacement of test specimen according to PGA. (**a**) X-direction; (**b**) Y-direction; (**c**) Z-direction.

Figure 18. Maximum strain of the bottom and bed beams of test specimen according to scaling factor. (**a**) At the bottom beam; (**b**) At the bed beam.

4. Operational Test of the Specimen

For operational capacity assessment, the mold transformer was tested before and after the series of shaking table tests. The tests were performed based on the International Electrotechnical Commission (IEC) IEC 60076-11 Standard [26] for dry-type transformers. Table 5 summarizes the test results of the transformer before and after the shaking table test series. From the test results, the specimen remained in good condition after the shaking table tests, in terms of external appearance. Moreover, the specification data of the transformer satisfied the acceptance criteria specified in IEC 60076-11 after the shaking table tests, thus ensuring the operational capacity of the test specimen. Nevertheless, the aforementioned weak points, including the loosening of linked bolts and the failure of spacers, can affect the operational capacity of the transformer in earthquakes with higher acceleration amplitude.

Table 5. Operational test results.

Name of Test	Acceptance Criterion (IEC 60076-11)	Test Results	
		Before Shaking Table Tests	After Shaking Table Tests
External appearance evaluation		Good	Good
Winding resistance measurement (at 100 °C) (Ω; Ω)	HV winding resistance LV winding resistance	0.174 0.000846	0.179 0.000797
No-load current (%)	<2.5	0.24	0.24
No-load loss (W)		1476	1487.00
Impedance voltage test (V; %)	<10	328.6 4.99	360.00 5.46
Load loss (W)		4859.66	5327.99
Full-load loss (W)		6335.66	6814.99
Efficiency (%)	>98.6	99.37	99.32
Voltage fluctuation rate (%)	<1.3	0.61	0.68
Insulation resistance measurement		Good	Good
Separate-source withstand voltage test		Good	Good
Impulse withstand voltage test		Good	Good
Temperature-rise test		Good	Good
Noise level test (dB)	<70	53.69	50.9
Partial discharge test	<10	5	5

5. Conclusions

In this study, the seismic performance of the electrical mold transformer was experimentally investigated through tri-axial shaking table tests. The input acceleration time histories were artificially generated according to the ICC-ES AC156 code with a range of different amplitude. A total of 22 shakings were performed during the entire test campaign. Based on the test results, the primary conclusions may be drawn as follows.

1. The dynamic properties of the test specimen were estimated through dynamic identification tests using random input signal. The initial natural frequencies of the mold transformer in the X, Y and Z-directions were 7.78, 2.52, and 23.12 Hz, respectively. Since the damage grew as the input motion amplitude increased, the fundamental frequency showed a decreasing trend; however, the level of frequency deterioration was not significant.

2. The damping ratios of the test specimen were evaluated using the half-power bandwidth method. The initial damping ratio of the test specimen was in the range of 2–4% and showed an increasing trend up to 4–10% with the increase of input ground motion amplitude.

3. In terms of damage stages, overall, at the final test, the specimen neither overturned nor collapsed; however, the spacers located on the bottom region of the transformer showed severe slippage. Simultaneously, the bolts connecting the bottom beam and bed beam were completely loose. Nonetheless, it should be noted that the prototype specimen was tested with conventional

anchoring details; thus, further investigations on the seismic performance of the mold transformers with different anchoring details should be taken into account to assess the damage characteristics of the mold transformer.

4. The dynamic amplification of the mold transformer by means of the acceleration amplification factor was evaluated. The amplification factors during the tests were in the range of 1.0–2.5 in the X, Y-directions, which complied with the ASCE 7-16 and FEMA E-74 Standard. Meanwhile, the acceleration amplification factors in the Z-direction were smaller than the lower limit specified in FEMA E-74.

5. During the shaking table test series, the maximum relative displacements in the X-direction did not exceed the boundary limit of 75 mm, which was recommended by the Korean National Radio Research Agency [25]. Meanwhile, the maximum relative displacements in the Y and Z-directions exceeded the boundary limit of 75 mm around the PGA of 0.50 and 0.47× g, respectively. Moreover, at the end of the shaking table tests, the maximum strain of the bottom beam and bed beam did not exceed the yield strain.

6. The specimen showed good condition of external appearance and satisfied the acceptance criteria specified in IEC 60076-11 after shaking table tests, thus ensuring the operational capacity of the transformer.

Author Contributions: Methodology, N.H.D.; Investigation, N.H.D., S.-J.L. and J.-Y.K.; Writing—original draft Preparation, N.H.D.; Writing—review and editing, K.-K.C.; Supervision, K.-K.C. All authors have read and agreed to the published version of the manuscript.

Funding: This research was supported by a grant (19AUDP-C146352-02) from the Architecture and Urban Development Research Program, funded by the Ministry of Land, Infrastructure and Transport of the Korean Government.

Acknowledgments: We gratefully acknowledge the supports of Ministry of Land, Infrastructure and Transport of the Korean Government, and Soongsil University.

References

1. Federal Emergency Management Agency (FEMA). *Reducing the Risks of Nonstructural Earthquake Damage: A Practical Guide (FEMA E-74)*; FEMA: Washington, DC, USA, 2011.

2. Whittaker, A.S.; Soong, T.T. An overview of nonstructural components research at three US Earthquake Engineering Research Centers. In *Proc. ATC Seminar on Seismic Design, Performance, and Retrofit of Nonstructural Components in Critical Facilities*; Earthquake Engineering Research Centers: Irvine, CA, USA, 2003.

3. Taghavi, S.; Miranda, E. *Response Assessment of Nonstructural Building Elements*; Pacific Earthquake Engineering Research Center: Irvine, CA, USA, 2003.

4. Di Sarno, L.; Magliulo, G.; D'Angela, D.; Cosenza, E. Experimental assessment of the seismic performance of hospital cabinets using shake table testing. *Earthq. Eng. Struct. Dyn.* **2019**, *48*, 103–123. [CrossRef]

5. Villaverde, R. Methods to assess the seismic collapse capacity of building structures: State of the art. *J. Struct. Eng.* **2017**, *133*, 57–66. [CrossRef]

6. Korea Architectural Institute. *Pohang Earthquake Damage Survey Report*; Korea Architectural Institute: Seoul, Korea, 2018.

7. Petrone, C.; Magliulo, G.; Manfredi, G. Shake table tests for the seismic assessment of hollow brick internal partitions. *Eng. Struct.* **2014**, *72*, 203–214. [CrossRef]

8. Fiorino, L.; Bucciero, B.; Landolfo, R. Evaluation of seismic dynamic behaviour of drywall partitions, façades and ceilings through shake table testing. *Eng. Struct.* **2019**, *180*, 103–123. [CrossRef]

9. Wang, S.J.; Yang, Y.H.; Lin, F.R.; Jeng, J.W.; Hwang, J.S. Experimental study on seismic performance of mechanical/electrical equipment with vibration isolation systems. *J. Earthq. Eng.* **2017**, *21*, 439–460. [CrossRef]

10. Hwang, H.H.; Huo, J.R. Seismic fragility analysis of electric substation equipment and structures. *Probabilistic Eng. Mech.* **1998**, *13*, 107–116. [CrossRef]

11. Porter, K.; Johnson, G.; Sheppard, R.; Bachman, R. Fragility of mechanical, electrical, and plumbing equipment. *Earthq. Spectra* **2010**, *26*, 451–472. [CrossRef]

12. Fathali, S.; Filiatrault, A. *Experimental Seismic-Performance Evaluation of Isolation/Restraint Systems for Mechanical Equipment Part I: Heavy Equipment Study Technical Report, No. MCEER-07-0007*; University at Buffalo, The State University of New York: Buffalo, NY, USA, 2007.

13. Fathali, S.; Filiatrault, A. *Experimental Seismic Performance Evaluation of Isolation/Restraint Systems for Mechanical Equipment Part 2: Light Equipment Study Technical Report, No. MCEER-07-0022*; University at Buffalo, The State University of New York: Buffalo, NY, USA, 2007.

14. International Code Council Evaluation Services Acceptance Criteria (ICC-ES AC156). *Acceptance Criteria for Seismic Certification by Shake-Table Testing of Nonstructural Components*; International Code Council Evaluation Service, Inc.: Whittier, CA, USA, 2000.

15. Federal Emergency Management Agency (FEMA). *Interim Protocols for Determining Seismic Performance Characteristics of Structural and Nonstructural Components through Laboratory Testing (FEMA 461)*; FEMA: Washington, DC, USA, 2017.

16. Architectural Institute of Korea. *Korean Building Code*; Architectural Institute of Korea: Seoul, Korea, 2006.

17. European Committee for Standardization. *Eurocode 8: Design of Structures for Earthquake Resistance—Part 1: General Rules, Seismic Actions and Rules for Buildings, EN 1998–1*; European Committee for Standardization: Brussels, Belgium, 2011.

18. Wheeler, A.J.; Ganji, A.R. *Introduction to Engineering Experimentation*; Prentice Hall: Upper Saddle River, NJ, USA, 1996; pp. 180–184.

19. Ljung, L. *System Identification: Theory for the User*; Prentice Hall: Upper Saddle River, NJ, USA, 1999.

20. Irvine, T. The Half Power Bandwidth Method for Damping Calculation. 2005. Available online: www.vibrationdata.com (accessed on 17 June 2019).

21. Wang, J.; Lü, D.; Jin, F.; Zhang, C. Accuracy of the half-power bandwidth method with a third-order correction for estimating damping in multi-DOF systems. *Earthq. Eng. Eng. Vib.* **2013**, *12*, 33–38. [CrossRef]

22. Clough, R.W.; Penzien, J. *Dynamics of Structures*; Computers and Structures Inc.: Berkeley, CA, USA, 2003.

23. Legrand Group. Cast Resin Transformers. For Distribution, Rectification, Traction and Special Solution. 2016. Available online: https://www.legrand.nl (accessed on 17 June 2019).

24. American Society of Civil Engineers (ASCE). *Minimum Design Loads and Associated Criteria for Buildings and Other Structures*; ASCE Standard ASCE/SEI 7-16; ASCE: Reston, VA, USA, 2016.

25. Korean National Radio Research Agency. *Earthquake Resistance Test. Method for Telecommunication Facilities*; Korean National Radio Research Agency: Naju-si, Korea, 2015.

26. International Electrotechnical Commission (IEC). *Power Transformers-Part. 11: Dry-Type Transformers*; IEC Standard IEC 60076-11; IEC: Geneva, Switzerland, 2018.

Article

Seismic Response Analysis of Multi-Story Steel Frames Using BRB and SCB Hybrid Bracing System

Rong Chen [1,*], Canxing Qiu [2,*] and Dongxue Hao [1]

[1] School of Civil Engineering and Architecture, Northeast Electric Power University, Jilin 132012, China; 20102291@neepu.edu.cn

[2] Key Laboratory of Urban Security and Disaster Engineering of Ministry of Education, Beijing University of Technology, Beijing 100124, China

* Correspondence: 20112384@neepu.edu (R.C.); qiucanxing@bjut.edu.cn (C.Q.)

Received: 8 November 2019; Accepted: 23 December 2019; Published: 30 December 2019

Featured Application: The proposed bracing system provides a novel solution for controlling peak seismic responses and for eliminating residual deformation for conventional concentrically braced frames.

Abstract: Multi-story steel frames are popular building structures. For those with insufficient seismic resistance, their seismic capacity can be improved by installing buckling-restrained braces (BRBs), which is known for high energy dissipation capacity, and the corresponding frame is denoted as BRB frame (BRBF). However, BRBFs are frequently criticized because of excessive residual deformations after earthquakes, which impede the post-event repairing work and immediate occupancy. Meanwhile, self-centering braces (SCBs), which were invented with a particular purpose of eliminating residual deformation for the protected structures, underwent fast development in recent years. However, the damping capability of SCBs is relatively small because their hysteresis is characterized by a flag shape. Therefore, this paper aims to combine these two different braces to form a hybrid bracing system. A total of four combinations are proposed to seek an optimal solution. The multi-story steel frames installed with BRBs, SCBs, and combined braces are numerically investigated through nonlinear static and dynamic analyses. Interested seismic response parameters refer to the maximum story drift ratios, maximum floor accelerations, and residual story drift ratios. The seismic analysis results indicate that the frames using the combined bracing system are able to take the advantages of BRBs and SCBs.

Keywords: multi-story steel frames; self-centering bracing elements; buckling-restrained brace (BRB); seismic analysis

---

1. Introduction

Conventional multi-story steel frames were found susceptible to earthquake attacks, and the huge social and economic loss caused by catastrophic earthquakes inspired the community to explore advanced technologies to upgrade the seismic resistance of structures. Many seismic damping devices, such as those based on friction mechanism [1–3], metallic yielding behavior [4,5], and buckling-restrained braces (BRBs) [6], are representatives among many research efforts and have been applied in practice to protect the structures.

Initially, normal steel braces were installed in steel frames to enhance seismic capacity. For example, Shen et al. [7] compared normal brace and BRBs and indicated that BRB is more efficient than normal braces in earthquake resistance. Hsiao et al. [8] proposed a sophisticated model to capture the buckling behavior of normal steel braces. Simpson and Mahin [9] suggested using a strongback braced frame to mitigate buckling-induced weak story behavior. Being different from normal steel braces, BRBs avoid buckling-induced instability and thus show stable cyclic behavior under both tension and compression loads. Owing to their outstanding damping capacity and economical production, BRB frames (BRBFs) have become popular seismic resistant structural systems and are being widely used over many countries [10]. According to prior work [6,10], BRBs exhibited plumb hysteresis, which enables BRBFs to successfully absorb seismic energy. Sabelli et al. [6] numerically quantified the seismic demands of BRBFs. Ariyaratana and Fahnestock [11] evaluated the effect of reserve strength on the seismic performance of BRBFs. Many numerical studies can be found elsewhere [12]. However, it is noted that the plumb cyclic behavior of BRBs leads to excessive residual deformation in the global system. As reported, under earthquakes, BRBFs usually exhibited satisfactory seismic performance but were prone to producing noticeable residual deformation, even after moderate earthquakes [6,13]. Through the numerical analysis reported by Sabelli et al. [6], the average residual story drift ratio of multi-story BRBFs was over 0.5% when subjected to design basis earthquake (DBE) ground motion records.

Excessive residual deformations may hinder the post-event recovery of building function and repairing work and, thus, have gained increasing attention among the community of earthquake engineering [14–16]. Post-event investigations found that a lot of building structures were finally demolished and rebuilt due to unrecoverable deformation, although they performed satisfactorily during earthquakes. Recognizing the problems associated with excessive residual deformations, the community quantitatively defined several thresholds of residual deformation that correspond to several damage levels. When the residual story drift ratio was over 0.5%, it would be more suggestible to rebuild a new structure rather than repairing the damaged one [17]. Besides, with the reparability, the out-of-plumb limits were also determined by the magnitudes of residual story drift ratios, which were determined to be 0.2% and 0.1% for the 3- and 6-story frames, respectively, based on relevant analytical results [18].

Many strategies were recently proposed and found effective in reducing the residual deformation demands of BRBFs [19–26]. For example, adopting moment-resisting frame (MRF) as the backup system for BRBFs was a methodology recommended by Reference [27] from the aspect of system level. By paralleling MRF to share 25% of the total design base shear, the residual story drift ratios of BRBF was reduced remarkably although not completely eliminated [27]. Ariyaratana and Fahnestock [11] also found that the participation of MRF reduced the residual story drift ratios. In their work, the residual deformations were decreased by approximately 50%. Combining MRF brings noticeable efficacy on reducing residual story drifts; however, it deserves further improvement because the residual story drifts were still over 0.5% upon a few earthquake records that correspond to the DBE level. As a consequence, the BRBF-MRF system still violated the reparability limit defined by a post-earthquake renaissance report [17].

As have been pointed out by prior studies [22,26,28–38], self-centering braces (SCBs) are an alternative to conventional braces because they can recover large deformation and can absorb input seismic energy. SCBs can be based on shape memory alloys (SMAs) or post-tension technology supplemented by damping sources, exhibiting a flag-shape (FS) hysteresis. The first SMA-based SCB can be traced back to the work by Dolce et al. [29]. In their study, the SMA-based SCBs consisted of a pair of superelastic SMA cables which were prestressed to a desirable extent to provide recentering capacity and energy dissipation ability, respectively, and then they were tested within a scaled reinforce concrete frame. Christopoulos et al. [30] invented an innovative SCB which explored the large elastic behavior of aramid tendons and friction damper. According to the test, the SCB exhibited full FS behavior and suffered from high axial strain level while resulting in zero structural damage. Through extensive analyses on FS single-degree-of-freedom (SDOF) systems, Christopoulos et al. [39] pointed out the benefits of increasing energy dissipation and post-yield stiffness in improving the seismic performance of self-centering structures. Researchers have proposed various improvement schemes for systems using SCBs. Such as Zhu and Zhang [22] tuned the relative amount of the prestressed and un-prestressed SMA cables to meet the energy dissipation and post-yield stiffness of SCB at a desirable level. In another work, the authors [40] introduced friction mechanism in SMA-based SCBs to increase the damping behavior of self-centering hysteresis. The testing results of the reduced-scale brace specimen detected that the FS cyclic behaviors were repeatable for many loading cycles without any property degradation. Erochko et al. [41] explored the friction damper, in which the friction surfaces were created inside in the SCBs to generate a plump hysteresis. Further, the SCBs were tested within a reduce-scale frame on shaking table and they performed as expected in a series of strong earthquakes and maintained a stable FS hysteresis with high damping capacity.

As have been revealed by prior studies, BRBFs have the problem of generating excessive residual deformations after earthquakes, while the self-centering concentrically braced frames (SCCBFs) have relatively low energy dissipation capability. This work comprehensively compares the seismic behavior of BRBFs, SCCBFs, and concentrically braced frames (CBFs) using hybrid bracing systems. The idea of combing BRBs and SCBs in a braced frame has never been proposed by earlier studies. To validate the proposed idea, nonlinear static and time-history analyses were carried out on the bracing system using BRBs, SCBs, and a combination of them. In the comparative analysis, interested seismic response parameters include maximum story drift ratio, maximum floor acceleration, and residual story drift ratio. Through the comparative study, it indicates that the BRBFs are featured with excessive residual deformation and that the SCCBFs are characterized by larger deformation demand. However, when the BRBs and SCBs are utilized together as the hybrid bracing system, the protected frame structures respond with improved performance, including reduced maximum deformation, controlled maximum floor acceleration, and minimized residual deformation. As well as should be noted is that, although this paper primarily analyzed the seismic performance of multi-story CBFs, the associated observations and conclusions might also provide some insights into the similar self-centering structures and systems.

2. Cyclic Behaviors of BRB and SCB

Figure 1 shows that the cyclic behaviors of BRB and SCB are usually simplified as a bilinear elasto-plastic and FS constitute model, respectively. The simplified models have been either experimentally or numerically validated by many prior studies [6,13,15,20]. As suggested in prior studies [6,11,15], the post-yield stiffness ratio of BRBs is within the range of 1% to 5%. In this analysis, this value for BRB is defined to 1%. An idealized FS model is used to describe the cyclic behavior of SCB in this study. Besides with the "yielding" strength and elastic stiffness, the cyclic behavior of the FS model is determined by introducing another two critical parameters, e.g., the post-yield stiffness ratio α and hysteresis width β. In fact, the values of these two parameters may vary with the practical configuration of the bracing components and configurations. Through properly tuning the contribution of each individual subassembly member, the SCBs can achieve a number of combinations of α and β. To be consistent with the post-yield strength of BRB and to maximize the energy dissipation capacity, the values of α and β currently considered for SCBs are set to be 1% and 1.0, respectively. It is noted as well, to simplify the problem, the degradation mechanism associated with stiffness and strength is excluded.

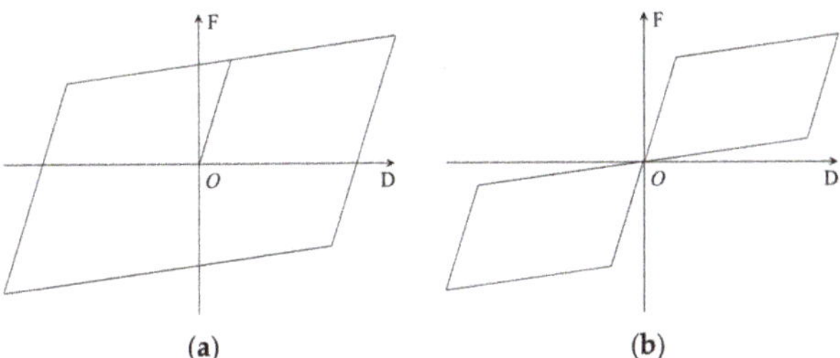

Figure 1. Cyclic behaviors of the considered braces: (**a**) buckling-restrained brace (BRB); (**b**) self-centering brace (SCB).

As aforementioned, SCBs are assumed to perform identically as BRBs in the initial elastic stage, which is the primary premise made in the comparative analysis. In such a way, the comparison is focused on highlighting the seismic response characteristics of CBFs when different types of braces are installed. It should be noted that the "yield" behavior of SCB is analogous to that of a normal steel brace but that the structural member does not really yield. The "yield" strength of a SCB actually refers to the stiffness degradation behavior of the component. In detail, for SCBs using post-tension technique, the "yield" behavior is caused by the decompression or activation force [30] while for those based on SMAs, it stands for the strength triggering the austenite to martensite phase transformation upon loadings [29]. In this analysis, the fracture problem of braces is not considered, which follows the assumption made by a variety of earlier studies [6,13–15].

3. Multi-Story CBF

Sabelli et al. [6] designed a six-story BRBF that complies with the requirement of National Earthquake Hazards Reduction Program (NEHRP) [42], assuming the location is Los Angeles downtown. This frame is used in the following analysis. The same parameters of BRBs and frame members are used to be consistent with the original design. Figure 2 shows that the CBF, which was noted as 6vb2 in original study, has a chevron-bracing placement. In this study, it is renamed as S1. It has a bay width of 9.0 m for each single bay, and its 1st-story height is 5.5 m, and the upper stories are 4 m. More building information and design procedure can be found in a related work by Sabelli et al. [6]. With the aim to isolate the influence of framing mechanism, the beam-to-column rigid connections are amended to hinge type. Such a treatment also helps to release bending moment suffered by the connection area and to adapt to large rotations while maintaining the connections damage free [43]. The connection amendment weakens the lateral stiffness of the system and thus leads to a longer structural vibration period to a certain degree. Incidentally, the story number is one parameter that needs in-depth investigations in the future. The corresponding effect can be quantified by increasing and decreasing the number of stories. However, the parametric analysis on story number is not included in this work due to space limitation and current research focus.

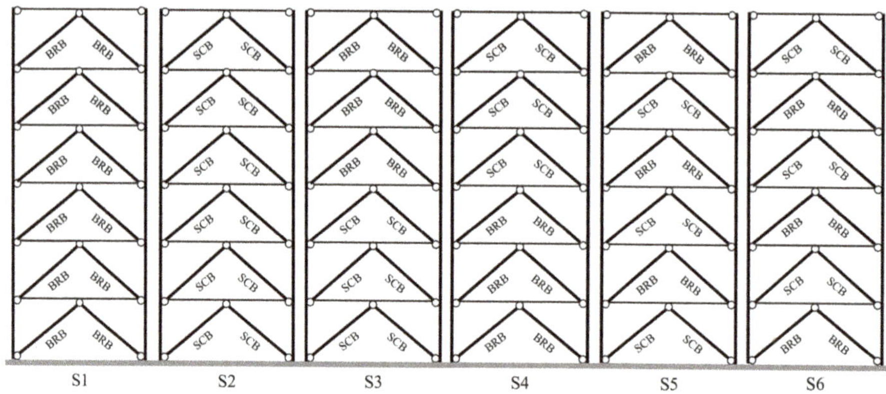

Figure 2. Concentrically braced frames (CBFs) using various bracing systems.

In terms of the other frames under consideration, they are produced by substituting the BRBs in frame S1 with SCBs at different locations. Figure 2 shows the notation of the frames. The BRBF is denoted as S1, and the identical SCCBF which uses SCBs throughout building height is named as S2. The effect of varying the combinations of BRBs and SCBs is intensively examined. The structures of S3 and S4 are achieved by replacing the BRBs with SCBs at the lower and upper three stories, respectively. For the structures of S5 and S6, the BRBs are replaced alternately throughout building height. Specifically, the BRBs are replaced with SCBs at the odd and even stories in S5 and S6, respectively. The yield strength F_y and the elastic stiffness k are defined identical for the BRBs and SCBs in every single story. Consequently, the framing structures, although they are installed with various bracing system, have exactly the same dynamic properties in the initial elastic state. According to the eigen value computation, these CBFs have a fundamental period of approximately 0.82 s. It is also worth noting that the performance objectives under earthquakes are not explicitly specified, since the current focus is paid on seeking a solution to reduce the residual deformation of BRBFs through comparing the frames installed with different bracing systems.

4. Numerical Model

A total of six CBFs numerical models are created in the earthquake simulation platform OpenSees [44]. Figure 3 displays the numerical model. Figure 3 shows the building information of the prototype frame as well, which presents the cross-sectional shapes of the main members and the building dimension. The designations of W14 × 48, W14 × 132 and W14 × 211 are the steel beams and columns with wide-flange or I-shaped sections. Taking W14 × 48 for example, it means the steel has a cross-sectional height of 14 in. with a unit weight of 48 lbs/ft. For all stories, Table 1 includes the strength capacity and the initial axial stiffness of the braces. The beams, columns, and braces are modeled using the force-based beam-column elements, considering the displacement interpolation functions of displacement-based elements may deviate from the exact solution [45]. Columns have fixed bases and are continuously modeled throughout building height. The beam-to-column pinned connections are achieved by using two over lapped nodes. The beam and column elements are made of American Society of Testing Materials (ASTM) A992 steel material. The corresponding post-yield stiffness ratio is assumed to be 0.003. The Rayleigh damping ratio of 5% is assumed for the first two modes. Similar to a prior treatment [46], the panel zone is not simulated, with the aim to keep the analysis time to a reasonable level. Further, to simplify the problem, the buckling and fatigue problems are excluded, and thus, the potential degradation of strength and stiffness of steel were not taken into account. In terms of the bracing components, each one is modeled as one single element with four integration points; the corresponding cross sections are an assembly of uniaxial fibers. Due to symmetry, only one bay is numerically modeled. Vertical gravity loads are applied gradually on the model, and then, dynamic analysis under horizontal seismic input is carried out. The structural torsion along with a vertical axis is not triggered.

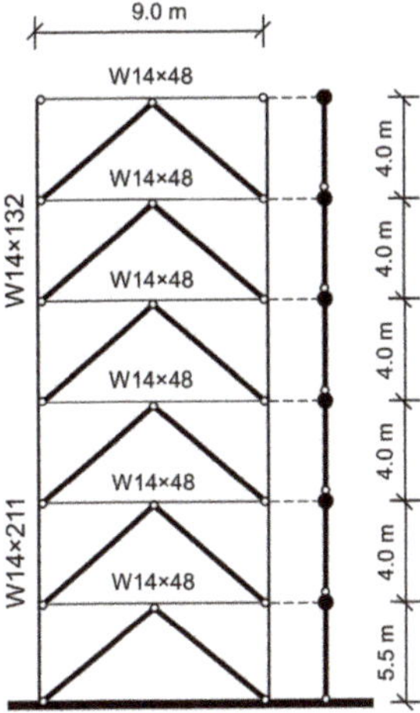

Figure 3. Modeling of the prototype frame in OpenSees.

Table 1. Mechanical properties of the braces.

Level	Yield Strength (kN)	Elastic Stiffness (kN/m)
1	1704	288×10^3
2	1306	244×10^3
3	1161	217×10^3
4	961	179×10^3
5	711	133×10^3
6	394	73×10^3

Figure 3 shows the tributary floor mass is idealized as concentrated mass nodes affiliated to the leaning column. The leaning column is to represent the other spans of frames that have negligible seismic resistance while carry all the building weight. The leaning column is coupled to the braced frame at each floor level in the vertical and lateral displacement directions. This column is vertically rigid and has a large cross-sectional area to carry the applied nodal mass. In the numerical model, the leaning columns are jointly connected at the adjacent levels. In such a situation, the leaning column would not provide stiffness and strength in the seismic input direction for the entire system. The total seismic mass at each floor is equally shared by the braced bays, thus each frame carries 1/6 of the total floor mass. The floor and total mass of each braced frame are 1.51×10^5 kg and 9.06×10^5 kg, respectively. As a result, the leaning column is able to account for the P-Δ effect while does not affect the strength or stiffness of the structure.

5. Ground Motions

An ensemble of earthquake ground motions containing records developed by Somerville et al. [47] is considered. There are 20 single records in this suite, which are generated for Los Angeles area having a 10% probability of exceedance over 50 years, corresponding to design-basis earthquake (DBE) hazard level. They are designated as LA01–LA20. These records cater to the soil type S_D. Figure 4a plots spectral accelerations for the 5%-damped SDOF systems under these ground motion records. Figure 4b includes the earthquake records associated with the maximum considered earthquake (MCE) hazard level. It is seen that the mean spectrum over 20 records matches the design basis earthquake spectrum reasonably well. The fundamental period of the frames (i.e., $T_1 = 0.82$ s) is of particular interest. The spectral accelerations corresponding to the fundamental period are extracted from the response spectra. The mean value of 20 individual results is 0.90 g, which agrees very well with the result from the ground motion record LA09. Thus, the seismic performance under this single record will be examined in case study.

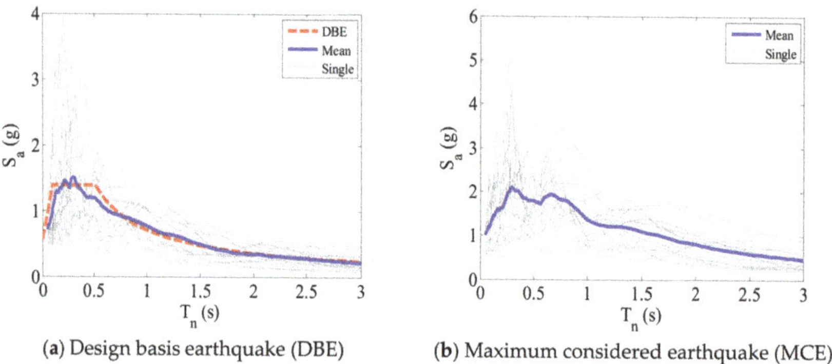

(a) Design basis earthquake (DBE) (b) Maximum considered earthquake (MCE)

Figure 4. Spectral acceleration of the 5%-damped single-degree-of-freedom (SDOF) system of the selected ground motion records.

6. Pushover Analysis

The braces are the kernel seismic resistant components of the braced frames. Figure 1 indicates that the BRB and SCB show different hysteretic behaviors upon a same cyclic loading loop. Accordingly, when installed with different braces, the CBFs demonstrate various nonlinear responses. Using the combined bracing system in the CBFs changes the global behavior of the framing systems, and the associated influence is firstly assessed by conducting the static pushover analysis. Gravity loads are gradually applied to the frame model prior to applying lateral forces. During the pushover procedure, the lateral force pattern compliant with the first vibration mode is applied and maintained. A control node is set at the roof level to monitor the target displacement. The roof drift ratio target is determined to 4%, which is sufficiently large to deform the structure into significant inelasticity.

Figure 5 assembles the cyclic behaviors of different framing systems by building the relationship between roof drift ratio and normalized base shear. The normalized base shear refers to the entire base shear divided by total building weight. As expected, the yielding strength, initial stiffness, and post-yield stiffness ratio are exactly the same among all the considered structures, since the elastic behavior of the braces is the same. However, the unloading behaviors are noticeably different. It is seen that the BRBF exhibited full elasto-plastic hysteresis as that of BRB components, showing a post-yield stiffness ratio of approximately 1.3%. In terms of the identical SCCBF, the hysteresis of the structural system is featured by purely FS behavior, which indicates the structural response is dominated by the local SCB component. Figure 5c–f plots the hysteresis shape of the CBFs using hybrid bracing systems, indicating the resulted cyclic behavior is between the bilinear elastic-plastic hysteresis and FS hysteresis. Compared with the BRBF, S3 to S6 tend to exhibit a pinching behavior due to the recentering property offered by SCBs. Compared with the SCCBF, the hysteresis is well widened, thanks to the high energy dissipation capacity provided by the BRBs. In other words, through combining BRBs and SCBs, the original BRBF gains improved recentering capability at the cost of losing energy dissipation capacity to a certain degree.

The assessment of global cyclic properties of different framing systems indicates the energy dissipation capacity of CBFs using hybrid braces lower than the BRBF. On the other hand, compared with the SCCBF, the maximum possible residual deformation, which usually refers to the intersection point of unloading line and deformation axis [48] is increased by introducing BRBs. However, the static analytical results do not necessarily reflect the dynamic properties under earthquakes. Thus, it is crucial to evaluate the effect of using a hybrid bracing system on the seismic behavior by nonlinear time history analysis.

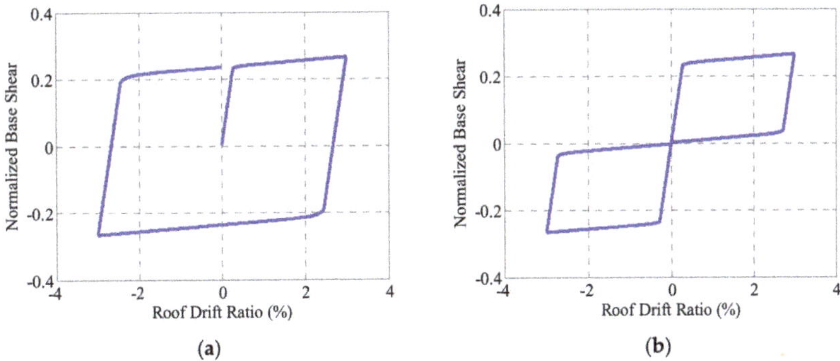

Figure 5. *Cont.*

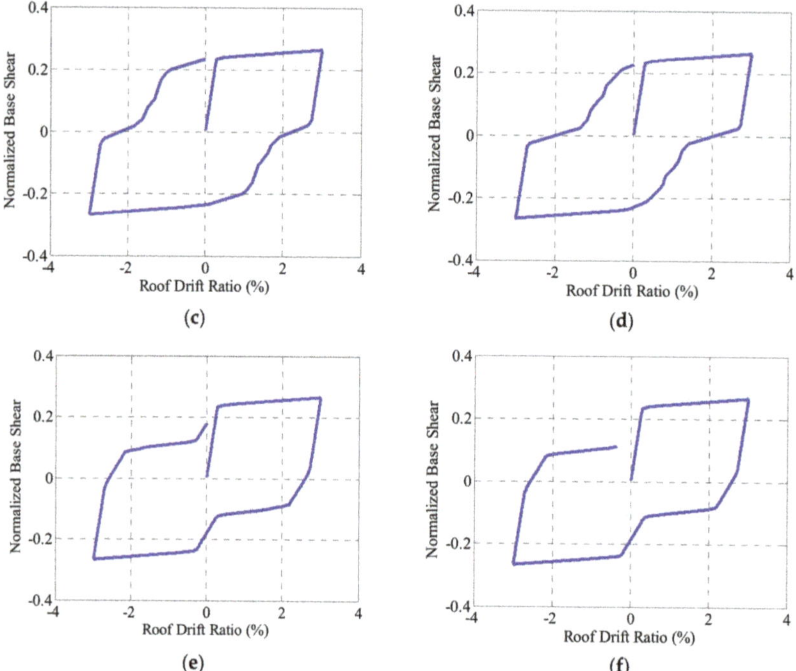

Figure 5. Cyclic pushover results of the steel frames: (**a**) S1; (**b**) S2; (**c**) S3; (**d**) S4; (**e**) S5; and (**f**) S6.

7. Nonlinear Time History Analysis

7.1. Case Study

To have a preliminary understanding of the seismic responses of the CBFs using hybrid bracing system, this subsection conducts nonlinear time history analysis on BRBF, SCCBF, and hybrid braced CBFs upon a single ground motion record by a case study. The ground motion record LA09 is artificially selected to demonstrate the effect of using hybrid bracing system. Figure 6 plots the time-history responses of roof drift ratios and roof accelerations for the frames under consideration. In the initial stage, the structures exhibit similar behavior until the ground motion excites the structures into serious nonlinearity. The peak roof drift ratios are approximately 0.66% and 1.05% for BRBF and SCCBF, respectively, which is primarily due to the fact that BRBF has much higher damping capacity than SCCBF. In terms of using hybrid braces, the corresponding deformation demands are between the BRBF and SCCBF, lying in the range from 0.91% to 1.01%. However, as will be shown in the statistical results, the average peak story drift ratio shows a different observation, which will be discussed later on.

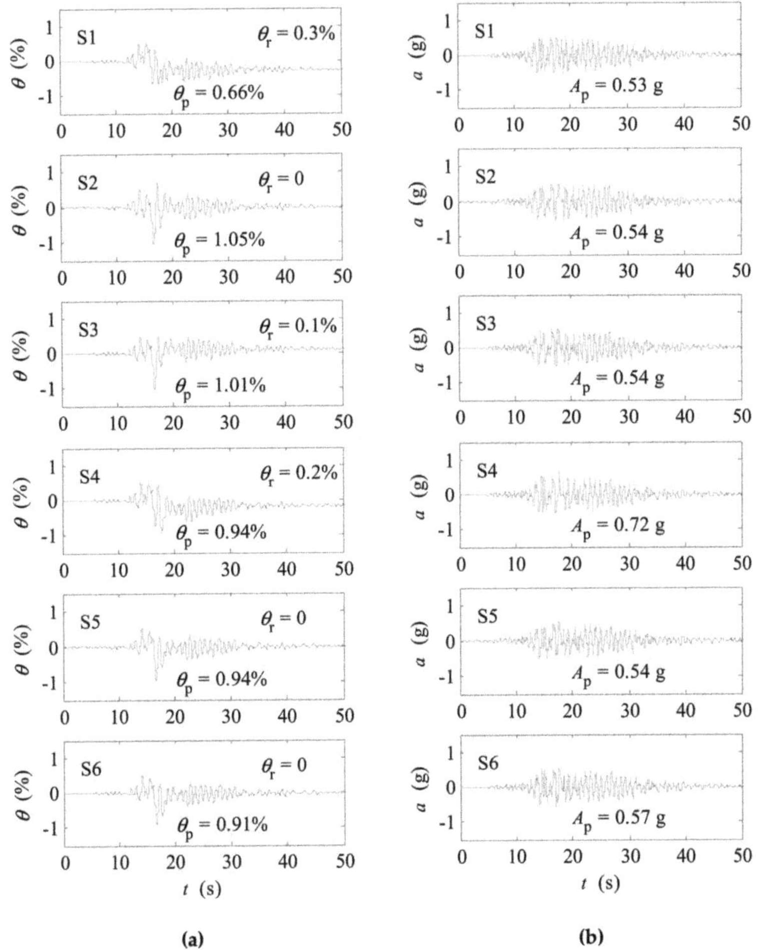

Figure 6. Response time history under ground motion LA09: (**a**) roof deformation; (**b**) roof acceleration.

The residual deformation demands are also a key index determining the seismic performance of seismic-resisting structures. This response parameter directly determines the repair cost and downtime duration after earthquakes. It is clear that the BRBF tends to produce biased deformation in a single side after being yielded, causing significant residual deformation at the end of the earthquake with 0.3% residual roof drift ratio. The SCCBF is found with entirely eliminated residual roof drift ratio at the end of vibration, thanks to the excellent recentering capability given by SCBs. Regarding the CBFs with hybrid bracing systems, the corresponding residual deformations are well reduced, compared with the BRBF. Therefore, the involvement of SCBs in a BRBF system may amplify the deformation demand to a certain degree but it remarkably decreases residual deformation demand.

Besides with the peak and residual deformation demands, the floor acceleration response is also examined. The results at the roof level are presented. It is interesting to note that, except for the frame S4, all the other frames suffer from nearly identical demands. This is because the yielding mechanism of the braces well caps the force demand in the entire structural system and consequently limits the increase of acceleration demand. The relatively large outcome generated in frame S4 is caused by higher mode effect. It means the significant nonlinearity of frame S4 involves the contribution from the 2nd or even higher modes. This phenomenon is similar to that noticed in a prior study [49].

Figure 7 plots the level-by-level responses throughout building height, including the peak story drift ratio, peak floor acceleration, and residual story drift ratio. This plot not only exhibits the magnitudes of considered demands but also displays the response deviation among different stories. Figure 7a shows that the SCCBF and BRBF exhibit similar peak deformation patterns over building height but that the former experiences higher demand by approximately 70%. When some of BRBs are replaced by SCBs, the deformation demands will be increased. Taking frames S5 and S6 for example, both of which use BRBs and SCBs alternatively over all stories, they show similar deformation patterns. Although the peak story drift ratios are approximately 50% larger than the BRBF, they are well within the design targets of 1.5% as prescribed by the American Society of Civil Engineers (ASCE) code [50]. For peak floor accelerations, the smallest values are associated with the BRBF. The amplified accelerations are mostly found in the floors where the BRBs are replaced, which is due to the low damping capability of SCBs. Among the four hybrid bracing configurations, the types of S5 and S6 are more favorable than the others because of their smaller demands. In terms of the residual story drift ratio, as expected, significantly residual deformations are found in the BRBF. At the second story, it is approximately 0.5%, implying the building is not suitable for repairing. However, the residual deformation demands are well reduced by installing SCBs, as evidenced by the results of SCCBF and hybrid bracing frames. It is also noted that structures S5 and S6 exhibit trivial demands over building height.

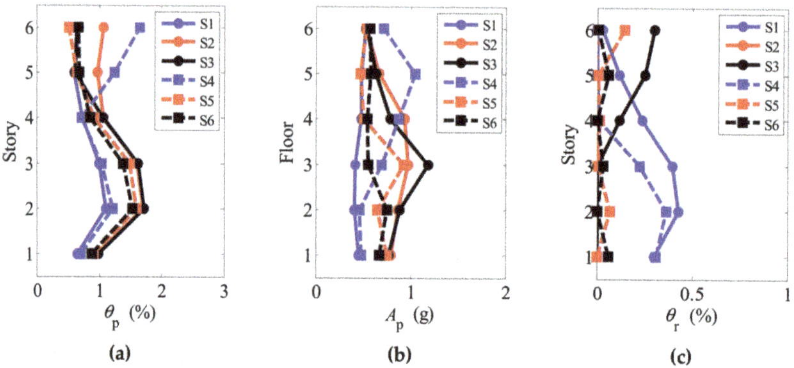

Figure 7. Seismic demands under ground motion LA09: (**a**) peak story drift ratio; (**b**) peak floor acceleration; (**c**) residual story drift ratio.

7.2. Statistical Results

As revealed by the spectral acceleration shown in Figure 4, remarkable record-to-record response deviation is noticed. Thus, the structural performance under an individual earthquake record does not necessarily stand for the central tendency when the structures are subject to a suite of ground motion records. Accordingly, the observation from the case study based on a single ground motion analysis is further examined through statistical analysis on all seismic analysis results. Figure 8 assembles the maximum story drift, maximum floor acceleration, and maximum residual story drift ratio under 20 ground motion records.

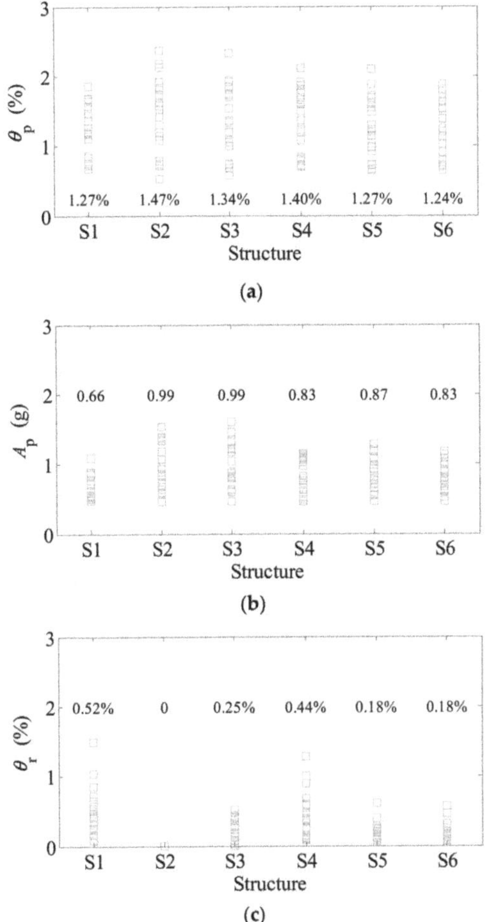

Figure 8. Seismic responses under the input ground motion records: (**a**) maximum transient story drift ratio; (**b**) maximum floor acceleration; and (**c**) maximum residual story drift ratio.

Figure 8a shows the scatter plot of the maximum story drift ratio for the structures under consideration. The mean values are calculated and put under the dots. Depending on the bracing configuration, the demands of hybrid braced CBFs can be larger or smaller than that of the pure BRBF. The mean values are 1.27% and 1.47% for the BRBF and the SCCBF, respectively. Although the equivalent damping ratio of SCCBF is half that of BRBF, the deformation demand is only approximately 16% higher. Compared with the BRBF, the demands of S5 and S6 are equivalent or even smaller. Thus, when the BRBs are replaced by SCBs at some stories, the corresponding frames are possible to generate improved deformation performance to a certain degree. For frames of S3 and S4, they exhibit larger deformations than original BRBF. This indicates that the BRBs and SCBs are suggested to be installed alternatively along building height.

Figure 8b presents the scattered results of maximum floor accelerations along with the mean values. The BRBF shows the smallest demands of 0.66 g, due to the best energy dissipation capacity. The frames of S2 and S3 show the largest demands among all the frames with a mean value of 0.99 g, attributed to the mild damping mechanism. For the frames of S4–S6, the acceleration demands are nearly identical, having middle values between that of BRBF and SCCBF. Thus, it can be anticipated that the engagement of SCBs would constantly increase floor acceleration demands.

Figure 8c assembles the residual story drift results. The comparison between S1 and S2 is sharp because using SCBs eliminates residual deformations for the structures. This is attributed to the excellent recentering capability of SCBs. Similar observations and explanations can be found in other studies [15,26,35–38]. The effect of combining SCBs on mitigating residual deformation is much more pronounced compared with the peak deformations and peak accelerations. As can be seen, the BRBF produces a mean residual story drift ratio over 0.5%, which violates the repairing threshold [17]. This is completely eliminated by replacing all BRBs with SCBs. For the other cases, the residual deformation is reduced to different extents by combining SCBs in different stories. The most remarkable reduction is found in the frames of S5 and S6. The corresponding reduction ratio is approximately 65%.

Figure 9 examined the story-by-story seismic performance by plotting the height-wise demands. For each story or floor level, the mean values are calculated and presented. The values in Figure 9 are slightly different from that in Figure 8 because the results in Figure 8 are the mean of the maximum transient responses over all levels, while that in Figure 9 are the mean of the maximum transient responses at the specific level. Consequently, the results in Figure 9 will be slightly smaller than that in Figure 8. Generally, a consistent observation can be made when recalling the scatter plots given by Figure 8. The mean values of peak story drift ratios, residual story drift ratios, and peak floor accelerations over the building height are plotted to examine the distribution uniformity of the deformation demands. The smallest peak story drifts tend to occur at the first story due to the increased story stiffness contributed by the fixed column bases. The deformation demands of S5 and S6 are smaller than the BRBF at the upper two stories. The mean results of residual story drift ratios are well reduced in almost each story by introducing the SCBs. The combined frame tends to exhibit a more uniform deformation demand over the building height than the pure BRBF. Regarding the peak floor acceleration, frames S5 and S6 are comparable to the BRBF.

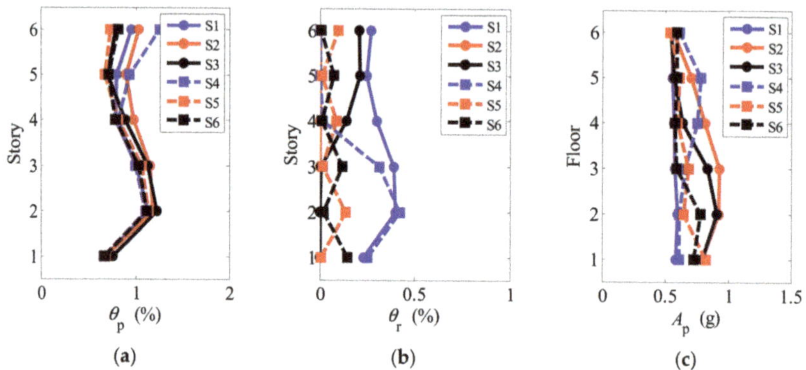

Figure 9. Seismic demands under the ground motion suite: (**a**) peak inter-story drift ratio; (**b**) residual inter-story drift ratio; and (**c**) peak floor accelerations.

Therefore, according to the assessment on the scattered maximum response and height-wise peak responses, it can be concluded that the frames of S5 and S6 use the most favorable hybrid bracing system among all the considered systems.

8. Conclusions

BRBFs tend to accumulate excessive residual deformation after earthquakes, which may prohibit the repairing work and cause long-term business downtime after earthquakes. On the other hand, SCBs are known for their excellent capability of recovering deformation. In order to address the problem for conventional BRBFs, this paper suggested using BRBs and SCBs together to form a hybrid bracing system. The proposed idea essentially combined the energy dissipation capacity of BRBs and the recentering capability of SCBs, which has never been reported in prior studies. The main assumption made in this analysis was that the fracture problem of braces could be avoided, which is also widely accepted in peer studies [6,13–15]. Thus, authors should understand current results with cautions. To validate the proposed methodology, intensive seismic analyses at the DBE seismic hazard level were conducted on six-story framing structures and the following conclusions can be obtained:

1. When the SCBs were installed to replace BRBs at certain stories, the global energy dissipation capacity will be deteriorated while the recentering capability was enhanced, according to the cyclic pushover results.
2. The hybrid bracing configurations that use SCBs and BRBs in alternative stories were suggested, considering such a placement reduced the maximum and residual story drift ratios by approximately 2% and 65%, respectively, compared with the pure BRBF.
3. Although the residual deformation cannot be completely eliminated by using the hybrid bracing system, it was well reduced to approximately 0.1%, which is small enough to carry out economical reparability.
4. The concept presented in the paper can shed light on the cases when different dissipative devices are used or when different structural typologies are considered.

Author Contributions: Conceptualization: R.C., D.H., and C.Q.; methodology: R.C. and C.Q.; software: R.C., C.Q., and D.H.; validation: R.C. and C.Q.; formal analysis: C.Q.; investigation: R.C. and C.Q.; resources: R.C., D.H., and C.Q.; data curation: D.H. and C.Q.; writing—original draft preparation: R.C. and D.H.; writing—review and editing: C.Q.; visualization: R.C., D.H., and C.Q.; supervision: R.C. and C.Q.; project administration: R.C. and D.H.; funding acquisition: R.C., D.H., and C.Q. All authors have read and agreed to the published version of the manuscript.

Funding: This research was funded by National Natural Science Foundation of China, grant number 51808317 and 51409045.

Conflicts of Interest: The authors declare no conflict of interest.

References

1. Downey, A.; Cao, L.; Laflamme, S.; Taylor, D.; Ricles, J. High capacity variable friction damper based on band brake technology. *Eng. Struct.* **2016**, *113*, 287–298. [CrossRef]
2. Downey, A.; Theisen, C.; Murphy, H.; Anastasi, N.; Laflamme, S. Cam-based passive variable friction device for structural control. *Eng. Struct.* **2019**, *188*, 430–439. [CrossRef]
3. Tajammolian, H.; Khoshnoudian, F.; Rezaei Rad, A.; Loghman, V. Seismic Fragility Assessment of Asymmetric Structures Supported on TCFP Bearings Subjected to Near-field Earthquakes. *Structures* **2018**, *13*, 66–78. [CrossRef]
4. Aghlara, R.; Tahir, M.M. A passive metallic damper with replaceable steel bar components for earthquake protection of structures. *Eng. Struct.* **2018**, *159*, 185–197. [CrossRef]
5. Rezaei Rad, A.; Banazadeh, M. Probabilistic risk-based performance evaluation of seismically base-isolated steel structures subjected to far-field earthquakes. *Buildings* **2018**, *8*, 128. [CrossRef]
6. Sabelli, R.; Mahin, S.; Chang, C. Seismic demands on steel braced frame buildings with buckling-restrained braces. *Eng. Struct.* **2003**, *25*, 655–666. [CrossRef]
7. Shen, J.; Seker, O.; Akbas, B.; Seker, P.; Momenzadeh, S.; Faytarouni, M. Seismic performance of concentrically braced frames with and without brace buckling. *Eng. Struct.* **2017**, *141*, 461–481. [CrossRef]
8. Hsiao, P.C.; Lehman, D.E.; Roeder, C.W. Improved analytical model for special concentrically braced frames. *J. Constr. Steel Res.* **2012**, *73*, 80–94. [CrossRef]
9. Simpson, B.G.; Mahin, S.A. Experimental and numerical investigation of strongback braced frame system to mitigate weak story behavior. *J. Struct. Eng.* **2017**, *144*, 04017211. [CrossRef]
10. Qu, B.; Liu, X.; Hou, H.; Qiu, C.; Hu, D. Testing of buckling-restrained braces with replaceable steel angle fuses. *J. Struct. Eng.* **2018**, *144*, 04018001. [CrossRef]
11. Ariyaratana, C.; Fahnestock, L.A. Evaluation of buckling-restrained braced frame seismic performance considering reserve strength. *Eng. Struct.* **2011**, *33*, 77–89. [CrossRef]

12. Uang, C.M.; Nakashima, M.; Tsai, K.C. Research and application of buckling-restrained braced frames. *Int. J. Steel Struct.* **2004**, *4*, 301–313.

13. Tremblay, R.; Lacerte, M.; Christopoulos, C. Seismic response of multistory buildings with self-centering energy dissipative steel braces. *J. Struct. Eng.* **2008**, *134*, 108–120. [CrossRef]

14. Deylami, A.; Mahdavipour, M.A. Probabilistic seismic demand assessment of residual drift for Buckling-Restrained Braced Frames as a dual system. *Struct. Saf.* **2016**, *58*, 31–39. [CrossRef]

15. Qiu, C.; Zhang, Y.; Li, H.; Qu, B.; Hou, H.; Tian, L. Seismic performance of concentrically braced frames with non-buckling braces: A comparative study. *Eng. Struct.* **2018**, *154*, 93–102. [CrossRef]

16. Chancellor, N.; Eatherton, M.; Roke, D.; Akbaş, T. Self-centering seismic lateral force resisting systems: High performance structures for the city of tomorrow. *Buildings* **2014**, *4*, 520–548. [CrossRef]

17. McCormick, J.; Aburano, H.; Ikenaga, M.; Nakashima, M. Permissible residual deformation levels for building structures considering both safety and human elements. In Proceedings of the 14th World Conference on Earthquake Engineering, Beijing, China, 12–17 October 2008.

18. Eatherton, M.R.; Hajjar, J.F. Residual drifts of self-centering systems including effects of ambient building resistance. *Earthq. Spectra* **2011**, *27*, 719–744. [CrossRef]

19. Ricles, J.M.; Sause, R.; Garlock, M.M.; Zhao, C. Post-tensioned seismic resistant connections for steel frames. *J. Struct. Eng.* **2001**, *127*, 113–121. [CrossRef]

20. Tian, L.; Qiu, C. Controlling Residual Drift in BRBFs by Combining SCCBFs in Parallel. *J. Perform. Constr. Fac.* **2018**, *32*, 04018047. [CrossRef]

21. Liu, Y.; Wang, H.; Qiu, C.; Zhao, X. Seismic Behavior of Superelastic Shape Memory Alloy Spring in Base Isolation System of Multi-Story Steel Frame. *Materials* **2019**, *12*, 997. [CrossRef]

22. Zhu, S.; Zhang, Y. Seismic analysis of concentrically braced frame systems with self-centering friction damping braces. *J. Struct. Eng.* **2008**, *134*, 121–131. [CrossRef]

23. Clayton, P.; Berman, J.; Lowes, L. Seismic design and performance of self-centering steel plate shear walls. *J. Struct. Eng.* **2012**, *138*, 22–30. [CrossRef]

24. Seo, J.; Kim, Y.; Hu, J. Pilot study for investigating the cyclic behavior of slit damper systems with recentering shape memory alloy (SMA) bending bars used for seismic restrainers. *Appl. Sci.* **2015**, *5*, 187–208. [CrossRef]

25. Seo, J.; Hu, J. Seismic response and performance evaluation of self-centering LRB isolators installed on the CBF building under NF ground motions. *Sustainability* **2016**, *8*, 109. [CrossRef]

26. Qiu, C.; Zhu, S. Shake table test and numerical study of self-centering steel frame with SMA braces. *Earthq. Eng. Struct. Dyn.* **2017**, *46*, 117–137. [CrossRef]

27. Kiggins, S.; Uang, C.M. Reducing residual drift of buckling-restrained braced frames as a dual system. *Eng. Struct.* **2006**, *28*, 1525–1532. [CrossRef]

28. Qiu, C.; Tian, L. Feasibility analysis of SMA-based damping devices for use in seismic isolation of low-rise frame buildings. *Int. J. Struct. Stab. Dyn.* **2018**, *18*, 1850087. [CrossRef]

29. Dolce, M.; Cardone, D.; Marnetto, R. Implementation and testing of passive control devices based on shape memory alloys. *Earthq. Eng. Struct. Dyn.* **2000**, *29*, 945–968. [CrossRef]

30. Christopoulos, C.; Tremblay, R.; Kim, H.J.; Lacerte, M. Self-centering energy dissipative bracing system for the seismic resistance of structures: Development and validation. *J. Struct. Eng.* **2008**, *134*, 96–107. [CrossRef]

31. Xu, L.H.; Fan, X.W.; Li, Z.X. Development and experimental verification of a pre-pressed spring self-centering energy dissipation brace. *Eng. Struct.* **2016**, *127*, 49–61. [CrossRef]

32. Xu, L.H.; Xie, X.S.; Yao, S.Q.; Li, Z.X. Hysteretic behavior and failure mechanism of an assembled self-centering brace. *B. Earthq. Eng.* **2019**, *17*, 3573–3592. [CrossRef]

33. Ghowsi, A.F.; Faqiri, A.; Sahoo, D.R. Numerical Study on Cyclic Response of Self-centering Steel Buckling-Restrained Braces. In *Recent Advances in Structural Engineering*; Rao, A., Ramanjaneyulu, K., Eds.; Springer: Singapore, 2018; pp. 589–598. [CrossRef]

34. Hou, H.; Li, H.; Qiu, C.; Zhang, Y. Effect of hysteretic properties of SMAs on seismic behavior of self-centering concentrically braced frames. *Struct. Control Hlth.* **2018**, *25*, e2110. [CrossRef]

35. Qiu, C.; Zhao, X.; Zhang, Y.; Hou, H. Robustness of Performance-Based Plastic Design Method for SMABFs. *Int. J. Steel Struct.* **2019**, *19*, 787–805. [CrossRef]

36. Qiu, C.; Zhang, Y.; Qi, J.; Li, H. Seismic behavior of properly designed CBFs equipped with NiTi SMA braces. *Smart Struct. Syst.* **2018**, *21*, 479–491.

37. Qiu, C.; Li, H.; Ji, K.; Hou, H.; Tian, L. Performance-based plastic design approach for multi-story self-centering concentrically braced frames using SMA braces. *Eng. Struct.* **2017**, *153*, 628–638. [CrossRef]

38. Qiu, C.X.; Zhu, S. Performance-based seismic design of self-centering steel frames with SMA-based braces. *Eng. Struct.* **2017**, *130*, 67–82. [CrossRef]

39. Christopoulos, C.; Filiatrault, A.; Folz, B. Seismic response of self-centring hysteretic SDOF systems. *Earthq. Eng. Struct. Dyn.* **2002**, *31*, 1131–1150. [CrossRef]

40. Zhang, Y.; Zhu, S. A shape memory alloy-based reusable hysteretic damper for seismic hazard mitigation. *Smart Mater. Struct.* **2007**, *16*, 1603. [CrossRef]

41. Erochko, J.; Christopoulos, C.; Tremblay, R.; Kim, H.J. Shake table testing and numerical simulation of a self-centering energy dissipative braced frame. *Earthq. Eng. Struct. Dyn.* **2013**, *42*, 1617–1635. [CrossRef]

42. Federal Emergency Management Agency. *NEHRP Recommended Provisions for Seismic Regulations for New Buildings and Other Structures*; Federal Emergency Management Agency: Washington, DC, USA, 1997.

43. Fahnestock, L.A.; Ricles, J.M.; Sause, R. Experimental evaluation of a large-scale buckling-restrained braced frame. *J. Struct. Eng.* **2007**, *133*, 1205–1214. [CrossRef]

44. OpenSees. *Open system for Earthquake Engineering Simulation (OpenSees)*; v 2.4.1; Pacific Earthquake Engineering Research Center: Berkeley, CA, USA, 2013.

45. Neuenhofer, A.; Filippou, F.C. Evaluation of nonlinear frame finite-element models. *J. Struct. Eng.* **1997**, *123*, 958–966. [CrossRef]

46. Erochko, J.; Christopoulos, C.; Tremblay, R.; Choi, H. Residual drift response of SMRFs and BRB frames in steel buildings designed according to ASCE 7-05. *J. Struct. Eng.* **2010**, *137*, 589–599. [CrossRef]

47. Sommerville, P. *Development of Ground Motion Time Histories for Phase 2 of the FEMA/SAC Steel Project*; SAC Background Document SAC/BD-91/04; SAC Joint Venture: Sacramento, CA, USA, 1997.

48. MacRae, G.A.; Kawashima, K. Post-earthquake residual displacements of bilinear oscillators. *Earthq. Eng. Struct. Dyn.* **1997**, *26*, 701–716. [CrossRef]

49. Qiu, C.X.; Zhu, S. High-mode effects on seismic performance of multi-story self-centering braced steel frames. *J. Constr. Steel Res.* **2016**, *119*, 133–143. [CrossRef]

50. American Society of Civil Engineers. *Minimum Design Loads for Buildings and Other Structures*; American Society of Civil Engineers: Reston, VA, USA, 2010.

Article

Equivalent Frame Model with a Decaying Nonlinear Moment-Curvature of Steel-Reinforced Concrete Joints

Isaac Montava *, Ramón Irles, Luis Estevan and Ismael Vives

Departamento de Ingeniería Civil, Universidad de Alicante, 03080 Sant Vicent del Raspeig, Spain; ramon.irles@ua.es (R.I.); luis.estevan@ua.es (L.E.); ismael.vives@gcloud.ua.es (I.V.)
* Correspondence: isaac.montava@ua.es; Tel.: +34-966-528-428

Received: 7 November 2019; Accepted: 12 December 2019; Published: 16 December 2019

Abstract: A numerical model for the analysis of frame structures that is capable of reproducing the behavior of reinforced concrete (RC) members and steel-reinforced concrete (SRC) members in all steps until collapse by simulating a reduced resistance capacity is presented in this work. Taking into account the solid models obtained in previous research that have been validated by experimental results, moment-curvature graphics were obtained in all steps: elastic, plastic, and post-critical to collapse. Beam models versus 3D models considerably simplified the calculation of frame structures and correctly described both the plastic and post-critical phases. The moment-curvature graph can be used in a simplified frame analysis, from post critical behavior to collapse.

Keywords: steel reinforced concrete; joint; post critical; moment-curvature; nonlinear; frame model

1. Introduction

When seismic load actions are considered in steel-reinforced concrete (SRC), structure failure occurs mostly at the joints. A joint can be reinforced with an embedded steel cross-section to absorb a huge amount of energy in order to prevent a structure from failing. The greater the ductility, the larger the energy absorption during an earthquake and the larger the deformation that can be achieved before collapse, thus reducing the risk of injury to the people occupying the building.

The leading references for experimental studies on this subject are those carried out by Wakabayasi [1], who in 1973, reported the behavior of SRC structures.

Chen et al. [2] investigated over 17 specimens with different steel cross-section solutions for concrete. They were composed of L or T steel cross-sections with reinforced concrete. Their force–displacement graphs are comparable to other numerical studies that have been carried out, such as those of Yan et al. [3], who analyzed the hysteretic curves and introduced the attenuation coefficient to represent the effects of seismic damage.

Chen et al. [4] conducted different studies on steel-reinforced concrete joints. The results show that SRC joints efficiently dissipated energy. The superposition method was able to very accurately estimate the joint strength. The research by Wilkinson and Hancock [5] concluded that, after carrying out flexion tests on Class 1 rectangular hollow sections, it is not possible to show proper rotation for plastic designs. They define the capacity of rotation (R) according to the curvature (χ) and the cross-section and its plastic curvature (χ_p), where M_p is the plastic moment and EI is the elastic rigidity of the cross-section. The expressions for R and χ are given as Equations (1) and (2):

$$R = \frac{\chi}{\chi_p} \tag{1}$$

$$\chi_p = \frac{M_p}{EI} \tag{2}$$

The behavior of the linear structural members under flexion during nonlinear calculations of the plastic region can be better understood with moment-curvature graphs.

In a moment-curvature graph, the rotation capacity can be represented by the distance between the point where it reaches the plastic moment of the cross-section for the first time and the intersection point between the horizontal branch and the unloading one.

The studies of Anastasiadis et al. [6] provide us with a better understanding of the relationship between the rotation capacity of a beam and its ductility. These authors studied the rotation capacity of steel wide-flange beams, their mechanical characteristics, and the collapse mechanisms inside and outside the web plane. Two different descriptions of ductility can be considered: one is the curving capacity of the section, the other is the rotation capacity between the front and rear sections of a beam member portion. In this way, classes of sections can be classified according to Eurocode 3 [7]: Class 1 is a plastic section, Class 2 is a compact section, Class 3 is a semi-compact section, and Class 4 is as slender section, which can be classified according to the ductility of the beam element, namely high ductility, medium ductility, and low ductility. The second class is the most suitable to guarantee stress redistribution capacity and energy absorption [8].

The nonlinear relationship between the moment and curvature can be experimentally obtained from the theoretical behavior of the section by using extensometric wires or by means of numerical nonlinear models validated by experimental tests.

The definition of ductility as the ratio between the collapse curvature and elastic curvature (quoted in Eurocode 2 [9], despite talking about rotation instead) is not appropriate. Curvature is the ratio of the difference between the unitary deformation of superior and inferior fibers divided by the height of the section. By considering the integration of these curvatures along the beam axis, movements can be obtained. The elastoplastic model allows us to accurately predict behavior in all load phases during the loading process: a former lineal phase regarding serviceability situations and a plastic phase that allows us to predict behavior close to collapse.

Two cross-sections with identical resistance characteristics (in elastic moment and collapse moment terms) can show a very different moment-curvature (M–χ), regardless of whether the section is fragile or ductile.

The collapse curvature is bigger when ductile behavior occurs [10]. When fragile, phenomena happen very quickly with no warning time, and compressed concrete loses its resistance to displacement crack propagation and low deformations. As collapse load values are reached, stresses correspond to the sections in which the plastic hinges needed for collapse in the simplified model appear; the final figure is reached simultaneously among collapsing sections, while the values of other sections still remain far from this figure.

When comparing different moment-load graphics on different joints in a frame, several plastification grades are noticeable given the stress redistribution in the plastic phase. Gioncu and Petcu [11] studied the rotation capacity of double T- and beam-column joints when looked at from a local plastic mechanism point of view. They wrote code algorithms to obtain the beam rotation capacity, the results of which were in line with experimental tests.

Nowadays, finite element method (FEM)-based software is capable of solving several different types of analyses with multiple applications in the engineering field, from simple linear analyses to nonlinear complex analyses [12]. To solve the relevant equations, nonlinear calculations require an incremental process with increasing loads starting from a value of zero.

In recent years, many scientific works [13–15] have attempted to analyze the behavior of SRC joints. Models with moment-curvature graphs have become a very interesting tool to simulate the complex behavior of large structures [16]. The article describes the experimental tests carried out to better understand the behavior of different models.

The experimental tests to obtain the moment-curvature graphs were made in previous investigations [17]. Several experimental results were obtained for RC and SRC models.

In Montava et al. [18], three solid numerical models were tested to understand their behavior and obtain the moment-curvature graphs required in the present research.

The solid finite elements were modeled and validated with experimental results until reliable models were obtained. Starting with the numerical solid model, a moment-curvature graph was created to extrapolate the characteristics of the analyzed models to a frame structure of prismatic pieces, and to perform a nonlinear calculation on a wireframe model [19]. The main goal was to generate a new simplified frame model by taking into account the moment-curvature graph whose behavior better fits the experimental results. Moment-curvature graph decadence allows for the loss of the resistance of reinforcing bars of reinforced concrete and their breakage to be simulated using nonlinear calculations with large displacements. With this new validated simplified model, the intention was to better understand the behavior of 2D gantries as opposed to a horizontal load to simulate seismic action with vertical loads [20]. The displacements obtained in the different tested prototypes and the absorbed energies were compared. The simulation was validated by starting with a 3D solid model of a beam with two simple supports at the ends and a load at the mid-span, which was simulated. This new bar model, with the moment-curvature graphs taken as the main data source, allowed for complex frames to be simulated. The improvement of the SRC was demonstrated, and the matching results were reinforced only at the joints. The 3D solid model is not a complex tool, but accurately describes plastic and post-critical behavior and can be used for further analysis.

Recent articles have attempted to simulate the decreasing plastic behavior until breakage of reinforced concrete structures using moment-curvature graphs [21].

The main objective of this investigation is to obtain a numerical model of a moment-curvature graph to be used in a simplified frame analysis, from post-critical behavior to collapse. References [17,18] have been used in order to validate the experimental data versus the numerical data.

2. Description

2.1. Description of the Model

The process of analyzing the model was highly nonlinear, and therefore, involved very complex calculations as two non-linearities intervened at the same time: the geometric nonlinearity and the nonlinearity of the material's behavior in terms of the stress–strain curve. The decreasing branch of the resistance to rupture was included in this nonlinearity.

To build the wireframe model, the module APDL of Ansys (Version 16.2 and 17.2, Company: Ansys Inc., Canonsburg, Pennsylvania, United States, 2015), whose research license is held at the Department of Civil Engineering of the University of Alicante (Alicante, Spain). The intention was to simulate three prototypes (see Table 1):

Table 1. Table summarizing the performed tests, RC: reinforced concrete, SRC: steel-reinforced concrete.

Prototype	Typology	Concrete Sections (mm × mm)	Rebar Reinforcement	Cross-Sections	Distance between Supports (mm)
P03	RC	300 × 250	4 ø 12	-	3300
P04	SRC	300 × 250	4 ø 12	HEB-100	3300
P05	RC	300 × 250	2 ø 16 + 2 ø 20	-	3300

The experimental tests are detailed in Montava et al. [17]. The first model, P03, aimed to describe reinforced concrete. The model P04 had the same reinforcement by including an HEB-100 cross-section in the central part (which covered a length of 2000 mm). The third model, P05, had a reinforced concrete beam capable of supporting a similar load to the P04 model, but without the steel section (Figure 1).

(a)   (b)

Figure 1. Prototypes P03 and P04 in the execution phase (**a**) and image of the tested prototypes P03 and P04 (**b**) [17].

The characteristics of the joint in the reinforced concrete structures with the embedded steel cross-sections were essential to understanding their behavior because it increased the rigidity and ductility at the most vulnerable point, mainly in order to withstand seismic activity.

As the applied deformations were progressively increased, the following states represented in Figure 2 could be distinguished.

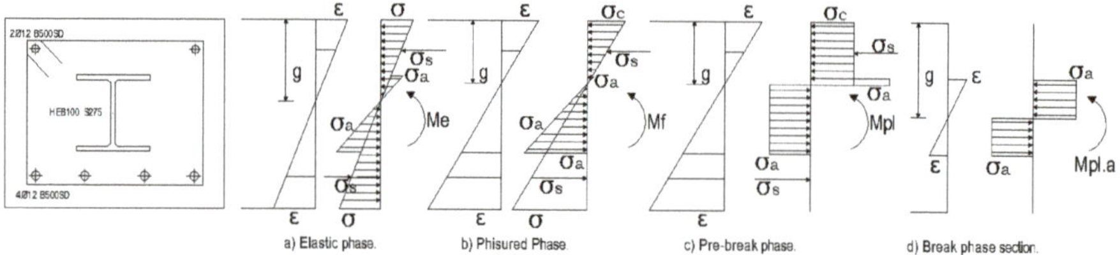

Figure 2. Tension state in different phases in a section of reinforced concrete with an embedded steel cross-section subject to simple bending.

2.2. Equivalent Frame Model with Ansys APDL M–χ Nonlinear Relationship

Obtaining the moment-curvature graphs allowed us to simulate wireframe structures using a nonlinear analysis because the moment of the section is related to its curvature; this is a relationship in which the reduced inertia of the concrete from cracking and loss of compression resistance, and even the rupture of bars, are implicit.

With this graph, data were input into the bar-type model using the BEAM188 (Ansys code) element, implemented with SECTYPE and GENB for non-linear calculations in bar sections (Figure 3).

$$
\begin{bmatrix} N \\ M_1 \\ M_2 \\ \tau \\ S_1 \\ S_2 \end{bmatrix} = \begin{bmatrix} AE(\varepsilon,T) & & & & & \\ & I_1E(\chi_1,T) & & & & \\ & & I_2E(\chi_2,T) & & & \\ & & & JG(\theta,T) & & \\ & & & & A_1G(v_1,T) & \\ & & & & & M_2G(v_1,T) \end{bmatrix} \begin{bmatrix} \varepsilon \\ \chi_1 \\ \chi_2 \\ \theta \\ v_1 \\ v_2 \end{bmatrix}
$$

N	Axial Force	A	Area	ε	Axial Strain
M_1	Bending Moment M_y	E	Modulus of elasticity	χ_1	Curvature K_{yy}
M_2	Bending Moment M_z	I2	Inertia	χ_2	Curvature K_{zz}
τ	Torsion	G	Modulus of shear	θ	Torsion curvature
S_1	Shear Force	A₁	Area	v_1	Transversal Shear Strain
S_2	Shear Force	T	Torsion	v_2	Transversal Shear Strain

Figure 3. Beam elements' non-linear behavior following a stress–strain relationship, matrix equation (Ansys help manual).

The simplified wireframe model allowed for a much simpler and more concise nonlinear calculation to be made than if the corresponding 3D solid model were used.

The behavior of beam elements is governed by the main moment-curvature relationship in the nonlinear calculation, and the other (axial, torsion, and shear) are described using linear relationships as their influence on the plastic behavior of the deflected bar is far weaker.

With greater ductility in the plastic phase, the structure allows for stresses to be better redistributed when some of its bars weaken, which is desired behavior for withstanding seismic actions.

3. Simulated Models

3.1. Prototypes P03, P04 and P05

The moment-curvature graph was obtained from the solid model [18]. It was made to compare the numerical model results with those obtained with the tested prototypes [17].

The novel procedure was used to build a 3D solid model of the finite elements validated with the experimental models and to obtain the moment-curvature graph from the deformations and moments obtained from the section.

It is important to note that for those materials showing deterioration, such as concrete, the moment-curvature graph changes to a decreasing trend for large curvatures, after verifying in this case that the behavior noted in the depletion of the modeled beam was similar to that tested if the graph changed to decreasing.

The moment-curvature graph was obtained from the solid model. The curvature obtained was the difference in the strain along the x-axis (beam direction) between the upper and lower nodes of the section 100 mm apart along the face of the column divided by the height of the section, see Figure 4.

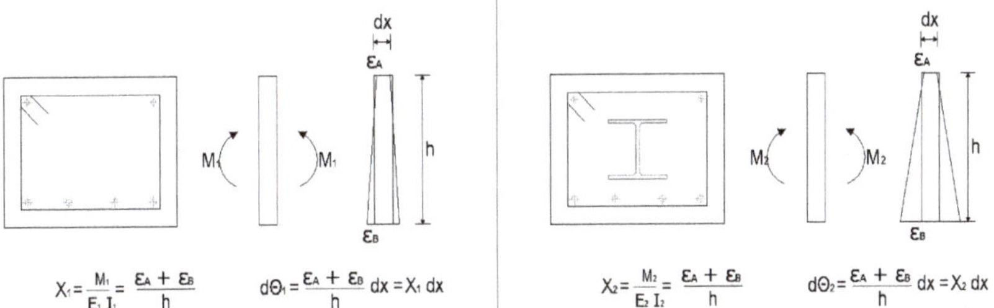

Figure 4. Analysis of the curvature of different sections.

The graph corresponding to prototype P04 allowed for a wider range of ductility compared to the other two obtained, where the break prevented greater deformations. It was possible to simulate the breakage of the section by including a drastic reduction that simulated the loss of resistance due to the deterioration of the concrete and breakage of reinforcements in the moment-curvature graph (Figure 5).

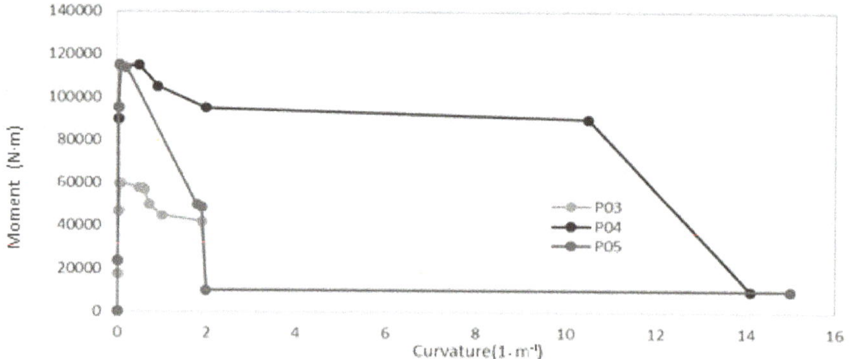

Figure 5. The moment-curvature graph of P03, P04, and P05 included in the frame model.

The moment-curvature graph was obtained from the solid model [18]. This novel procedure consisted of a three-dimensional finite element model validated by the experimental models to obtain the moment-curvature graph from the strains and moments in the section. The curvature obtained was the difference in the strain along the x-axis between the upper and lower nodes of the section 100 mm apart along the face of the column divided by the height of the section from the different prototypes P03, P04, and P05.

3.1.1. Reinforced Concrete P03 Prototype

From the moment-curvature graph, data were input into Ansys APDL to simulate a similar simplified bar model to the experimental test, and to verify its ability to reproduce the experimental tests and the corresponding 3D solid model. A displacement of 250 mm was introduced as the movement imposed to obtain the reactions and to verify that the behavior was similar to the experimental behavior (Figure 6).

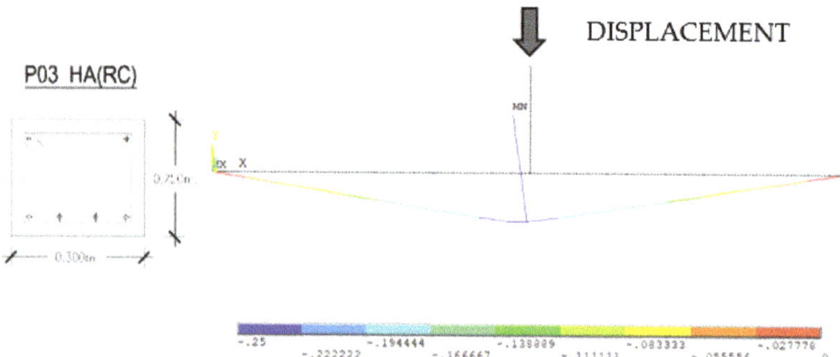

Figure 6. Vertical movement (m) of prototype P03 simulated according to the moment-curvature graph.

From Figure 7, in which the force–displacement graph is represented at the center of the beam, we can verify that the maximum load that the section resists was 74 kN in the FEM model versus 73 kN in the test, with a maximum displacement of 250 mm when the rupture of the reinforcing steel occurred.

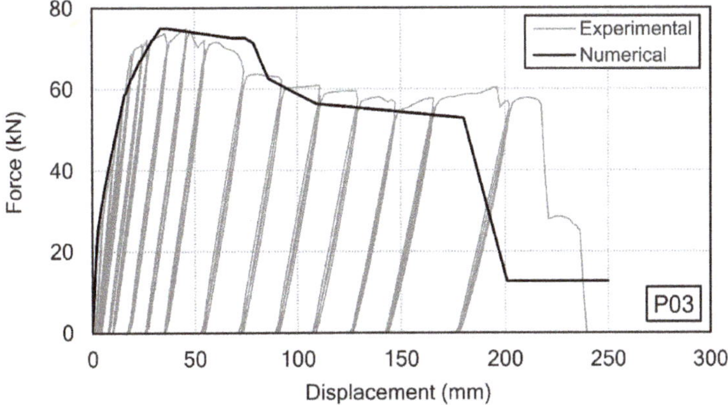

Figure 7. Force–displacement graph of the P03 model showing the results of the experimental tests and the numerical modeling of frames from the moment-curvature graph.

The behavior was elastoplastic, and it was possible to simulate a very complex behavior with significant deformations in both the concrete and steel from the nonlinear behavior of the moment-curvature graph (Figure 8). The reduced resistance of the calculation model was comparable to the experimental model.

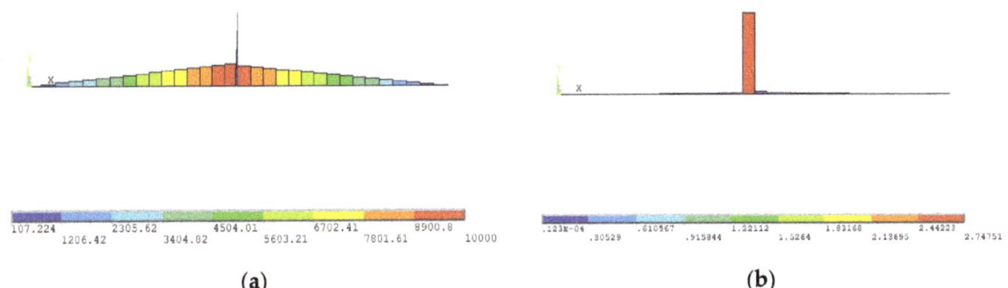

(a)   **(b)**

Figure 8. (a) Diagram of moments (N·m) and (b) curvatures (m^{-1}) during the maximum displacement.

3.1.2. Steel Reinforced Concrete P04 Prototype

The simulation was carried out with the frame model and the moment-curvature graph obtained from simulating the P04 prototype as in the previous section. In this case, the curvature that the section reached was much greater than in the previous one given the capacity of the steel cross-section's rotation in Montava et al. [17]; a 350 mm displacement was reached without having exhausted the steel cross-section and without considerably reducing the section's strength capacity (Figure 9).

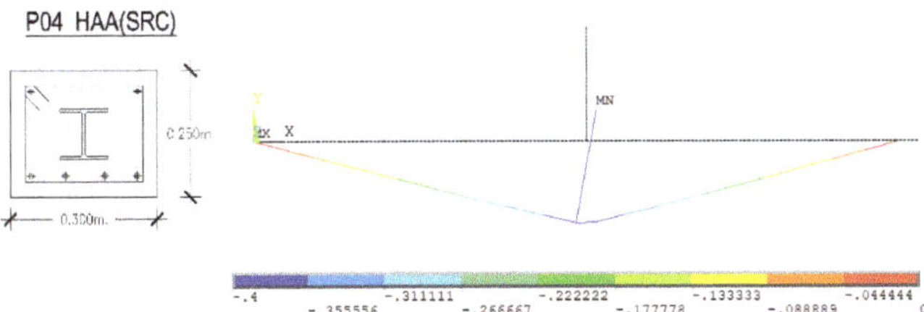

Figure 9.   Vertical movement (m) of prototype P04 simulated with the frame from the moment-curvature graph.

Figure 10 illustrates the diagram of moments (a) and curvatures (b) in the last step of a hinge with a high curvature in the section near the center.

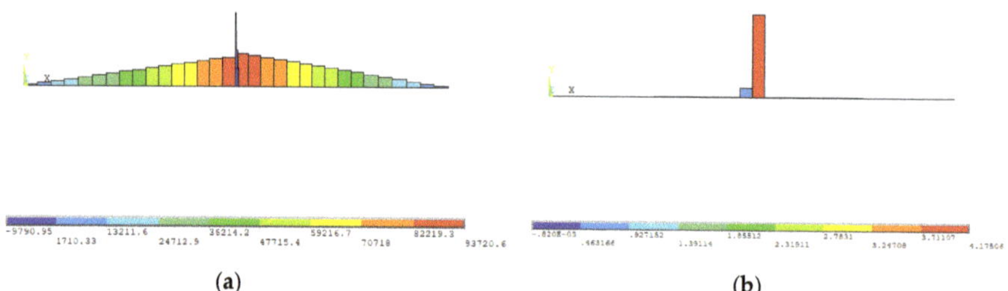

(a)   (b)

Figure 10. (**a**) Diagram of moments (N·m) and (**b**) curvatures (m^{-1}) during the maximum displacement.

The maximum load in the center of the beam that the section was capable of resisting was 150 kN in the simulated model, compared to 152 kN in the experimental test (Figure 11).

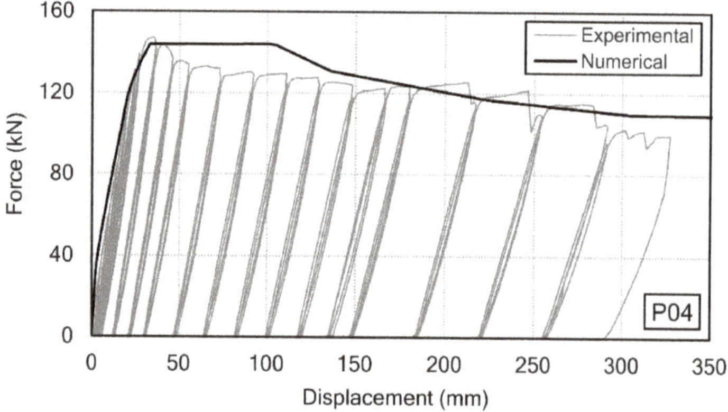

Figure 11. Force–displacement graph of model P04 showing the results from the experimental tests and the numerical model of frames from the moment-curvature graph.

3.1.3. Reinforced Concrete P05 Prototype

Prototype P05 was simulated to compare it with P04. It showed a similar resistance, but showed alower ductility than the steel-embedded cross-section. The maximum displacement was 250 mm (Figures 12 and 13).

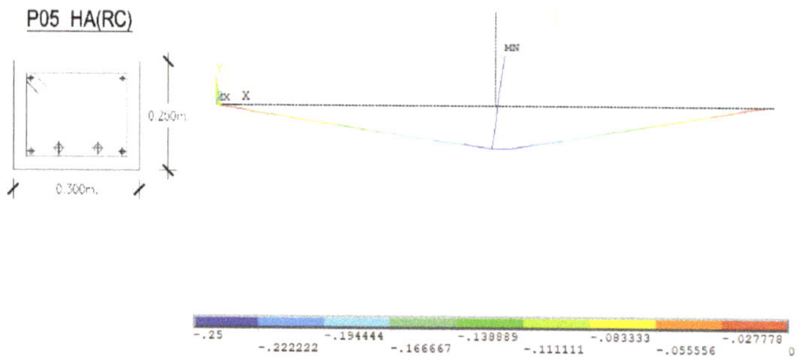

Figure 12. Vertical movement (m) of prototype P05 simulated from the moment-curvature graph.

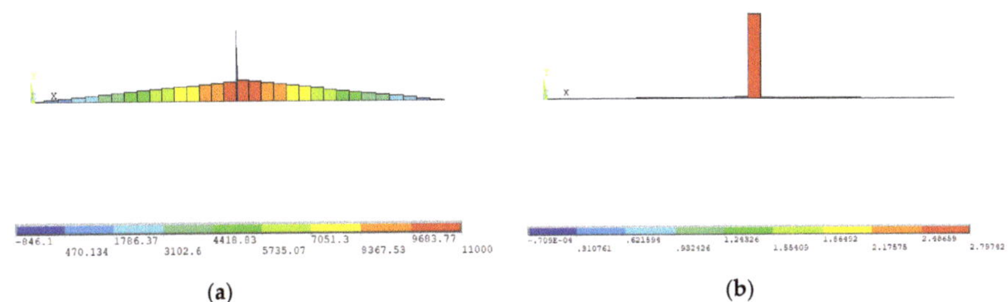

(a)   (b)

Figure 13. (**a**) Diagram of moments (N·m) and (**b**) curvatures (m^{-1}) during the maximum displacement.

The maximum load in the center of the beam that the section was capable of resisting was 145 kN in the simulated model, which was also 145 kN in the test (Figure 14).

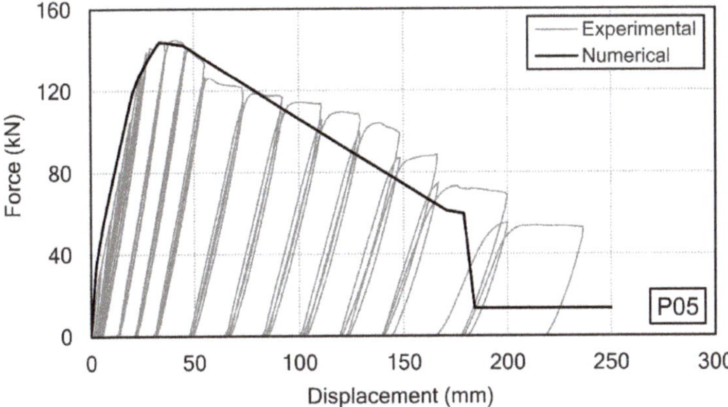

Figure 14. Force–displacement graph of model P05 showing the results of the experimental tests and numerical model of frames from the moment-curvature graph.

It was possible to simulate the decrease in the force–displacement graph. With 200 mm of displacement, this simulated section was in agreement with the moment-curvature graph obtained from the solid model.

3.2. Frame Structure in 2D

With the procedure validated in the two previous sections, we analyzed a classical frame structure of two columns and one beam to compare the maximum load, along with a horizontal displacement imposed on the top of the left column, that was able to be withstood by several arrangements of reinforcements.

To analyze the plastic behavior, an increasing horizontal displacement was imposed to collect data on the evolution of the variables during both plastification and the post-critical branch.

The beam was subjected to a vertical load in order to simulate the weight of the slab and the overload (10 kN/m) (Figure 15).

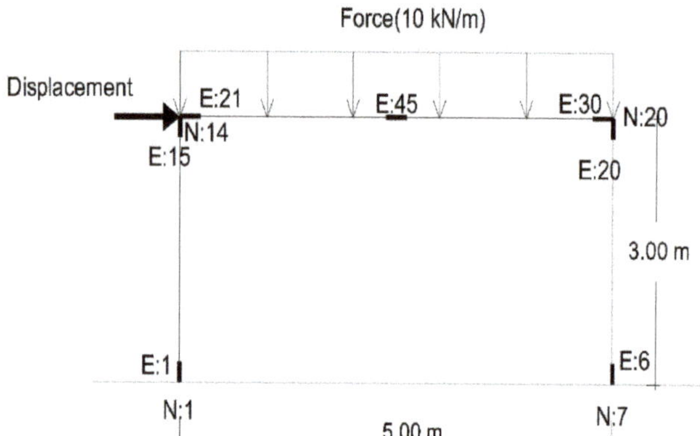

Figure 15. Frame model (N stands for node and E for element).

As it is a simplified frame model, the calculation was performed in a short time despite the nonlinearities (Table 2).

Table 2. Table summarizing the performed analyses.

Frame	Typology	Concrete Sections	Rebar Reinforcement	Cross-Sections
A	RC	300 × 250	4 ø 12	NONE
B	SRC	300 × 250	4 ø 12	HEB-100 (FULL FRAME)
C	RC	300 × 250	2 ø 16 + 2 ø 20	NONE
D	SRC + RC	300 × 250	4 ø 12	HEB-100 (JOINT ONLY)

3.2.1. Frame A (RC P03 Section)

Large displacements were applied to the joint until the maximum curvature value was reached at critical nodes, which was when the break took place (disintegration of concrete and breakage of reinforcements) (Figure 16).

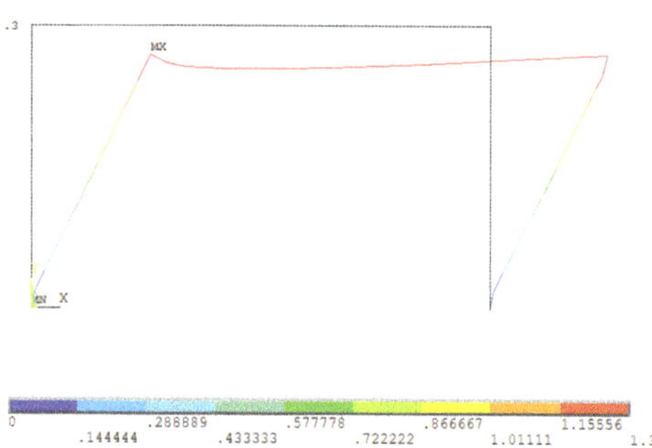

Figure 16. Horizontal movement (m) of the simulated frame structure from the moment-curvature graph of frame A.

The different nonlinear calculation steps correspond to points I, II, III, and IV in Figure 17.

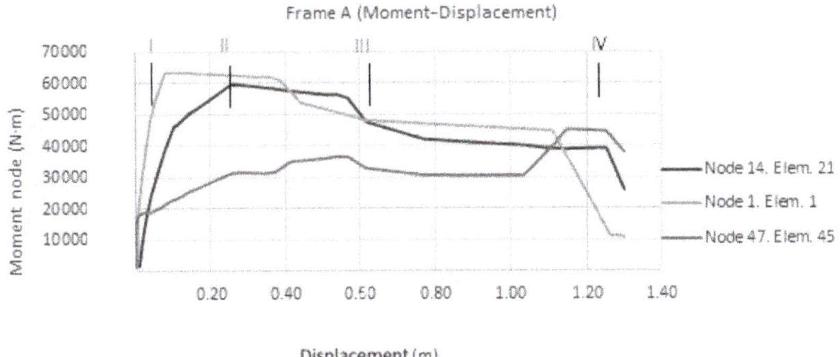

Figure 17. Graph of the moments of the different elements in relation to the displacement in node 14.

It was important to obtain the graph of the moments in the elements near the joints in relation to the displacement imposed on node 14 in order to compare the different behaviors of the nodes depending on the applied displacement and the degree of plastification. The graph of moments obtained (Figure 17) was based on the imposed displacement analyzed at node 14.

The collapse of the structure was simulated. In parts b, d, f, and h in Figure 18, the different curvatures of the modeled elements and their evolution with the imposed displacement can be seen.

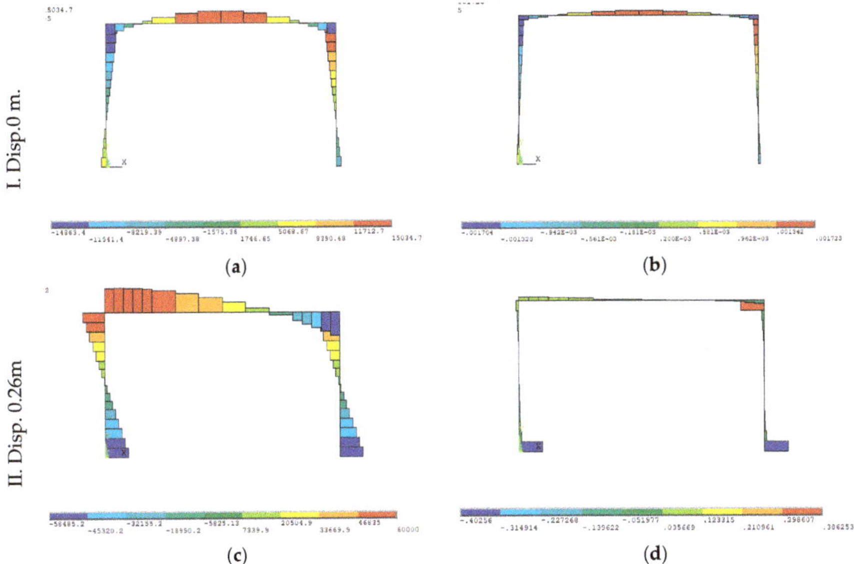

Figure 18. *Cont.*

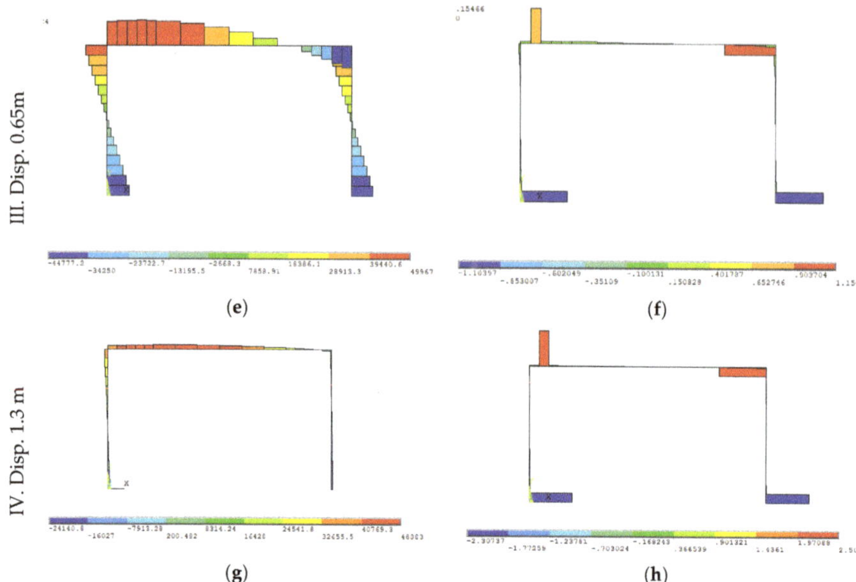

Figure 18. Diagram of the moments (N·m) in (**a,c,e,g**) and curvatures (m^{-1}) in (**b,d,f,h**) in the different displacements of node 14: I (0 m), II (0.26 m), III (0.65 m), and IV (1.3 m).

The appearance of hinges in the structure was seen with the very high curvature of the section. In Figure 18(I), the elements were found in the area of the proportional M–χ graph and had not reached the maximum moments. In Figure 18(II), some elements reached the maximum moment of the M–χ graph and began to plasticize. In Figure 18(III), some elements near the joint exceeded the maximum moment and the curvature increased considerably. Plasticization was intense, resulting in the development of high curvatures and hinge joints. The structure's strength thereby diminished. In Figure 18(IV), the values of stresses were very small when elements near the most plastified joint appeared with very high curvatures. They were considered to be plastified joints and appeared as a hinge. As this was a hyperstatic structure, different hinges appeared until the structure finally collapsed.

From the configuration that exhausted the elastic regime, the stresses were redistributed and displacements continued to increase considerably. Finally, the whole strength capacity collapsed.

3.2.2. Frame B (SRC P04 Section)

This was simulated with a horizontal 2-m displacement on the beam. Shifts higher than the previous model were reached because the boundary curvature of the SRC section was higher than that of reinforced concrete, which did not allow for such high curvatures (Figure 19).

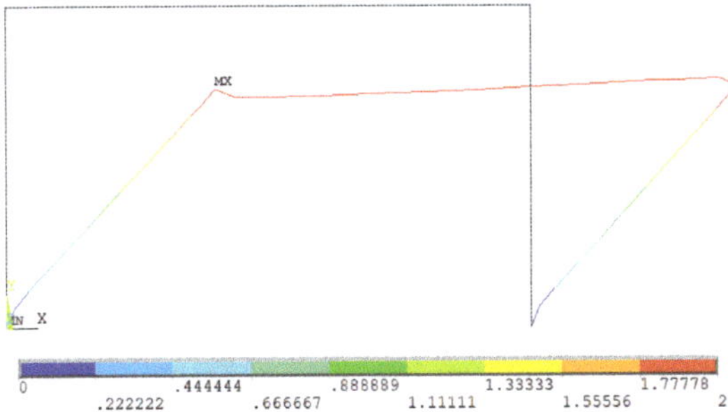

Figure 19. Horizontal movement (m) of the simulated frame structure from the moment-curvature graph of frame B.

Thus, the model and procedure to simulate the nonlinear calculation of reinforced concrete beams were validated (Figure 20). The maximum moments coincided with the maximum values of the moment-curvature graph, included to define the frame B section, which was 115 kN·m, and achieved for a 0.4 m displacement at node 14. Meanwhile, the maximum 2-m displacement at node 14 gradually reduced without reaching the maximum curvature of exhaustion. In the first calculation steps, the maximum moments were reached by increasing the curvature, and with it, displacements, and by redistributing the strength (Figure 21).

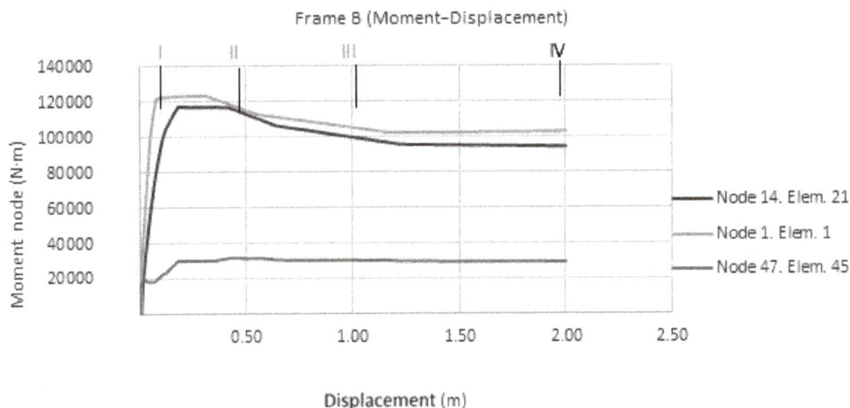

Figure 20. Graph of the moments in elements 1, 45, and 21 in relation to the displacement in node 14 of frame B.

It was verified that the capacity of the deformations was much greater than in the simulations modeled with the moment-curvature graph of reinforced concrete for frame A. This solution was able to resist large displacements before the structure collapsed by redistributing the forces and maintaining bearing capacity.

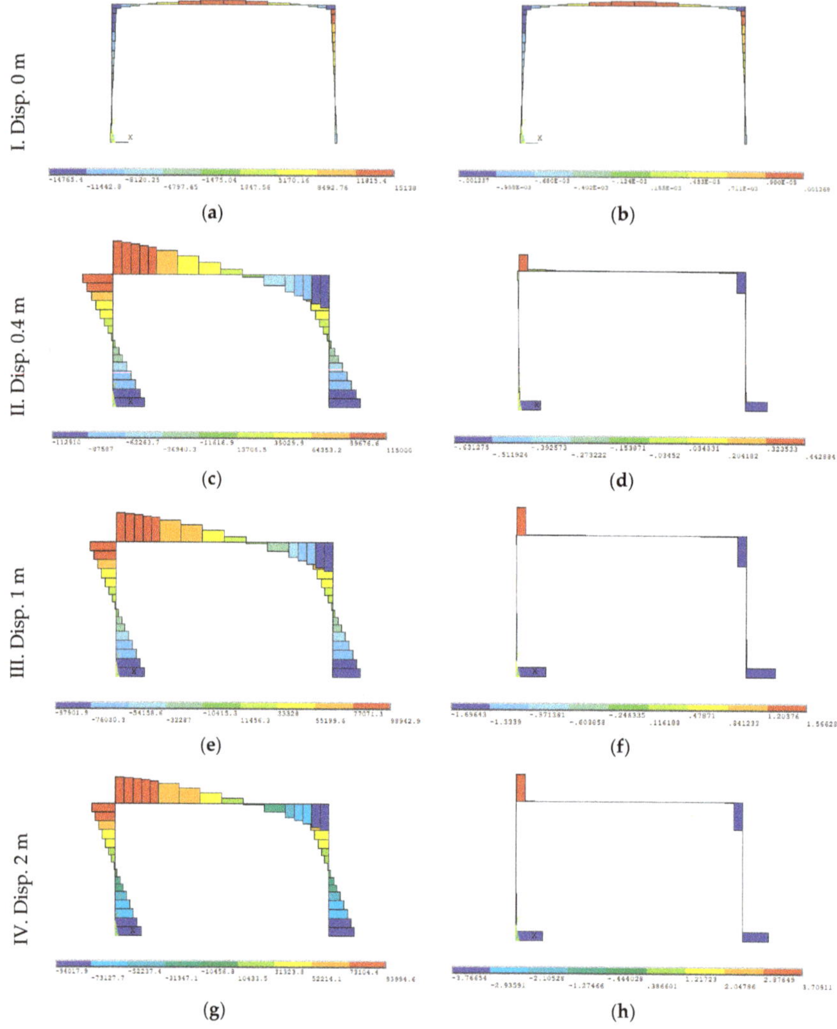

Figure 21. Diagram of the moments in (**a,c,e,g**) (N·m) and curvatures in (**b,d,f,h**) (m^{-1}) during the different displacements of node 14: I (0 m), II (0.4 m), III (1 m), and IV (2 m).

3.2.3. Frame C (RC P05 Section)

The two-column planar frame and one beam were simulated with a horizontal displacement in the lower 1.3 m joint with the moment-curvature graph of the prototype P05 section, which was the same strength in relation to the P04 prototype section but without the embedded steel cross-section.

The maximum moment that the frame structure modeled with the frame C graph was capable of resisting was 120 kN·m, which was equivalent to the frame structure modeled with the prototype P04 graph. During the maximum displacement imposed on node 14 (of 1.3 m) corresponding to the last step, a 30 kN·m moment was reached. This corresponded to the greater curvature values of the moment-curvature graph.

With the same moments, the maximum curvature that prototype P05 was capable of supporting was lower than that of prototype P04. When this curvature was surpassed, the tensioned steel bars exceeded their deformation limit and broke, which caused the structure to also break (Figures 22–24). The structure's ductility was considerably reduced.

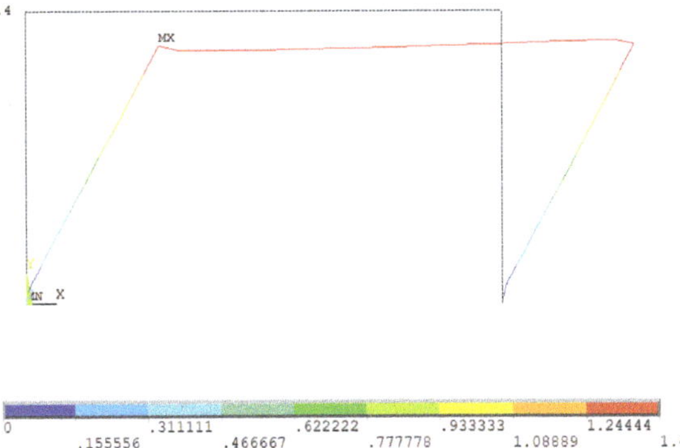

Figure 22. Horizontal movement (m) of the simulated frame structure from the moment-curvature graph of frame C.

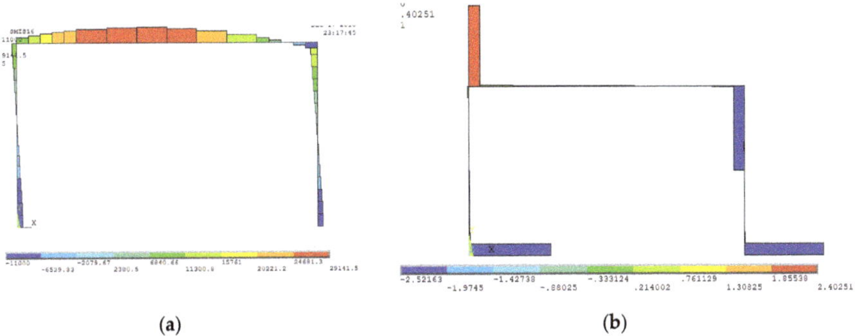

(a) (b)

Figure 23. (**a**) Diagram of moments (N·m) and (**b**) curvatures (m^{-1}) during the maximum displacement.

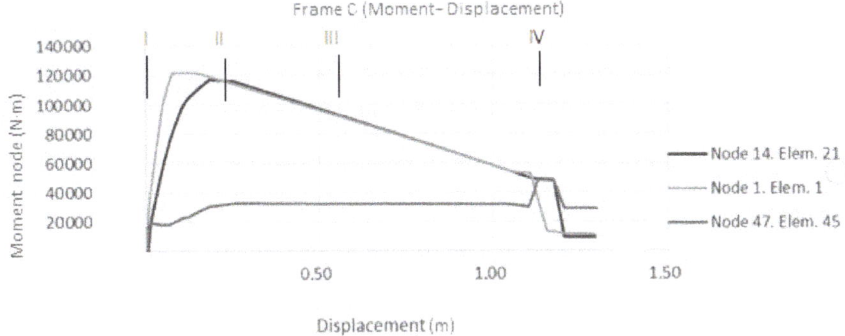

Figure 24. Graph of the moments in elements 1 and 24 front the displacement in Node 14 of frame C.

3.2.4. Frame D (RC P05 Section + SRC P03 Section Joint Only)

The same frame structure was simulated with the behavior of the moment-curvature graph of the reinforced concrete prototype section P05 in all the bars, except for joints, which were simulated from a fifth of the length at the ends of bars with the behavior of the moment-curvature graph of the SRC P04 prototype section.

The similarity of the results obtained with the simulated frame structure to the behavior of the moment-curvature graph of frame B, which extended to all the bars, was surprising. Its behavior was similar and offers consequent savings in material by incorporating steel cross-sections only into the joint. The structure's ductility was much greater than that of the previous model (Figure 25).

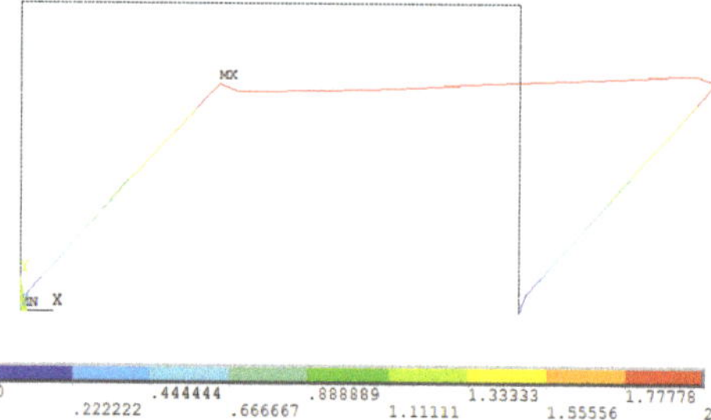

Figure 25. Horizontal movement (m) of the simulated frame structure from the moment-curvature graph of frame D.

3.2.5. Analysis of the Results

Figure 26 shows the results obtained from the different prototypes with the displacement applied at node 14 and the corresponding horizontal load. The highest energy absorption in frame B versus frame C is also observed in this figure. It resisted a comparable maximum value, but without the same ductility.

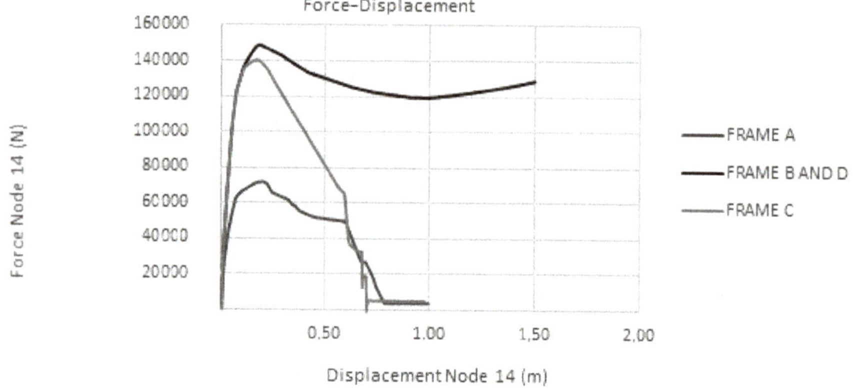

Figure 26. Graph of the load at Node 14 in relation to the displacement of the different frames.

The similarity of the curves in the frame B and frame D figures was remarkable given that the moment-curvature behavior of prototype frame B was simulated only in the joints to save material. Meanwhile, the rest of the bar was simulated with the behavior of frame C, which allowed for greater curvatures and bigger displacements than if joints were not reinforced, but with considerable savings in material.

The maximum moments coincide with the moments introduced from the moment-curvature graphs of each prototype. The simulated SRC frame achieved the greatest displacements, a bigger load capable of support, and was able to absorb the most energy. Therefore, it behaved better when faced with horizontal seismic loads, but showed no difference when we embedded the reinforcement of the steel cross-sections in only the joints, which is where the strength was greater and the plastification was concentrated.

The numerical model was obtained and a force–displacement graph from the simulated model with the moment–rotation graph was given. Once the non-linear simulation was carried out, the resulting force and the displacement of the most representative point of the gantry were obtained and the graph was drawn. When processing the non-linear calculation from the moment-curvature graph, a final result was obtained, from which, it was possible to extract the force–displacement values according to Figure 26. The force–displacement curve related to node 14 can be seen. The decreasing load starting from the maximum force the column was capable of, and the hardening at high displacements due to the strength mechanism change, with columns in tension are noticeable.

The results in the frame B and frame D graphs are the same.

The bending capacity of the steel cross-sections gave the structure a very high deformation capacity by maintaining major points of resistance until collapse took place due to a great energy absorption capacity and high ductility.

From the strengths (F) and displacement (D_{max}) of each 2D frame up until collapse, the energy absorbed by the structure can be calculated. The energy absorbed by the vertical force is calculated using Equation (3):

$$\text{Absorbed Energy} = \int_0^{D_{max}} F(u)\,du. \tag{3}$$

Table 3 summarizes the results obtained, from which we can conclude that the reinforced concrete (frame A) structures absorbed less energy than the steel-reinforced concrete structures (frame B). All the bars of the steel-reinforced concrete structures (frame B) coincided with the behavior of the model containing steel-reinforced concrete only at the joints of frame D. The latter was more advantageous as it offered the greatest absorption capacity of displacements and its respective strength, as well as an important saving in steel, which was achieved by using sections only at joints.

Table 3. Summary table of the performed tests.

	Frame A	Frame B	Frame C	Frame D
Displacement node 14 (m)	1.3	2	1.3	2
Force node 14 (kN)	75	160	160	160
Absorbed energy from the horizontal load (kN·m)	60	280	90	280

In the reinforced concrete structures, it would be convenient to explicitly consider the possibility of including reinforcements of reinforced concrete joints with embedded steel cross-sections as a solution for creating very high ductility according to Eurocode 4 [22] as the current regulation for designing structures with mixed steel and concrete.

Either a unidirectional or bi-directional reinforced concrete structure with flat beam slabs, which has a low ductility according to Eurocode 8 [23], can be improved by embedding steel cross-sections in the three spatial directions in all the structure's reinforced concrete joints to offer very high ductility. It would compete advantageously with the expensive construction devices that require high or very high ductility reinforced concrete with standard bars. The simplicity of the ordinary construction of flat floors when faced with seismic activity would be maintained.

The extension of Eurocode 4 [22] would be interesting for designing reinforced concrete structures with joints of fully embedded steel cross-sections, and not only in columns. The design can be achieved using moment-curvature graphs.

4. Conclusions

Based on the results of the numerical simulations obtained in this article, the following conclusions are drawn:

- Steel-reinforced concrete (SRC) structures double the ductility compared to the reinforced concrete (RC) structures, where a break prevented greater strains being reached.

- It was possible to simulate the elastoplastic or post-critical behavior until a section broke. A drastic reduction in the moment that simulated the deterioration of concrete, breakage of reinforcements, and loss resistance of steel cross-sections has been included in the moment-curvature graph.
- Hinges appeared simultaneously by redistributing forces along sections until collapse took place.
- SRC structures reinforced only at joints reduced the steel cross-section used in the structure compared to reinforcing all the bars with equal strength, ductility, and safety against a structure collapsing during earthquakes.
- Frame models with nonlinear moment-curvatures until the point of collapse were able to satisfactorily reproduce the behavior of three-dimensional models and the experimental test of the prototypes. The frame model was used to elucidate the hyperstatic behavior. It described the redistribution of strength and the structure's general behavior until a break occurred. It provides designers with a much simpler tool than the complete three-dimensional modeling of the contributed structure.
- It would be interesting to consider the generalization of the reinforcement with steel cross-sections embedded in the joint of reinforced concrete structures to achieve more earthquake-resistant structures, especially in public buildings and for emergencies where excellent seismic safety must be guaranteed.
- In the reinforced concrete structures, it would be convenient to explicitly consider the possibility of including reinforcements of reinforced concrete joints with embedded steel cross-sections as a solution for creating structures with very high ductility using Eurocode 4 as the current regulation for designing structures with mixed steel and concrete.
- The extension of Eurocode 4 would be interesting for designing reinforced concrete structures with joints having fully embedded steel cross-sections, and not only in columns, as contemplated by Japanese JIA(The Japan Institute of Architects) regulations. The design can be achieved by using moment-curvature graphs.
- Simulation with numerical models allowed for the analysis of complex situations. In particular, the model of simplified frames with the relationship of moment-curvature allowed for nonlinear calculations until large displacements were reached, taking into account the reduction of rigidity because of the cracking of the concrete.
- A new procedure was developed to obtain the moment-curvature graphs of the sections from the numerical models. The moment-curvature graph can be used in the simplified frame analysis by contemplating post-critical behavior in future research.
- The principal scientific contribution is that, using the Ansys program, it was possible to numerically validate a procedure to simulate until the breaking of a bar structure from some bar elements with a behavior introduced by means of the moment-curvature graph. The behavior was the same independent of whether all the bars were reinforced or only the joints were reinforced.

Author Contributions: Formal analysis, I.M., R.I., L.E. and I.V.; Investigation, I.M. and R.I.; Writing—original draft, I.M.; Writing—review & editing, I.M., R.I., L.E. and I.V.

Funding: This research received no external funding.

Conflicts of Interest: The authors declare no conflict of interest.

Data Availability: The data used to support the findings of this study were supplied by Isaac Montava under license and so cannot be made freely available. Requests for access to these data should be made to Isaac at isaac.montava@ua.es.

References

1. Wakabayashi, M. *Design of Earthquake—Resistant Buildings*; McGraw: New York, NY, USA, 1986.
2. Chen, C.C.; Lin, K.T. Behavior and strength of steel reinforced concrete beam–Column joints with two-side force inputs. *J. Constr. Steel Res.* **2009**, *65*, 641–649. [CrossRef]

3. Yan, C.; Yang, D.; Ma, Z.J.; Jia, J. Hysteretic model of SRUHSC column and SRC beam joints considering damage effects. *Mater. Struct.* **2017**, *50*, 88. [CrossRef]

4. Chen, Z.; Xu, J.; Xue, J. Hysteretic behavior of special shaped columns composed of steel and reinforced concrete (SRC). *Earthquake Eng. Eng. Vibr.* **2015**, *14*, 329–345. [CrossRef]

5. Wilkinson, T.; Hancock, G.J. Compact or Class 1 Limits for Rectangular Hollow Sections in Bending. In Proceedings of the 8th International Symposium on Tubular Structures, Singapore, 26–28 August 1998.

6. Gioncu, V.; Mosoarca, M.; Anastasiadis, A. Prediction of available rotation capacity and ductility of wide-flange beams. Part 1: DUCTROT-M computer program. *J. Constr. Steel Res.* **2012**, *69*, 8–19. [CrossRef]

7. European Committee for Standardization. *EN 1993-1-8, Eurocode 3: Design of Steel Structures—Part 1–8: Design of Joints*; European Committee for Standardization: Brussels, Belgium, 2005.

8. Li, J.; Wang, Y.; Lu, Z.; Li, J. Experimental Study and Numerical Simulation of a Laminated Reinforced Concrete Shear Wall with a Vertical Seam. *Appl. Sci.* **2017**, *7*, 6299. [CrossRef]

9. European Committee for Standardization. *EN 1992-1-1, Eurocode 2: Design of Concrete Structures—Part 1-1: General Rules and Rules for Buildings*; European Committee for Standardization: Brussels, Belgium, 2014; p. 730.

10. Lu, Z.; He, X.D.; Zhou, Y. Studies on damping behavior of vertically mixed structures with upper steel and lower concrete substructures. *Struct. Des. Tall Spec. Build.* **2017**, *26*. [CrossRef]

11. Gioncu, V.; Petcu, D. Available Rotation Capacity of Wide-Flange Beams and Beam-Columns. Part 1. Theoretical Approaches. *J. Constr. Steel Res.* **1997**, *43*, 161–217. [CrossRef]

12. Bossio, A.; Fabbrocino, F.; Lignola, G.P.; Prota, A.; Manfredi, G. Design Oriented Model for the Assessment of T-Shaped Beam-Column Joints in Reinforced Concrete Frames. *Buildings* **2017**, *7*, 118. [CrossRef]

13. Bossio, A.; Fabbrocino, F.; Lignola, G.P.; Prota, A.; Manfredi, G. Simplified Model for Strengthening Design of Beam–Column Internal Joints in Reinforced Concrete Frames. *Polymers* **2015**, *7*, 1732–1754. [CrossRef]

14. Santarsiero, G. FE Modelling of the Seismic Behavior of Wide Beam-Column Joints Strengthened with CFRP Systems. *Buildings* **2018**, *8*, 31. [CrossRef]

15. Kabir, M.R.; Alam, M.S.; Said, A.M.; Ayad, A. Performance of Hybrid Reinforced Concrete Beam Column Joint: A Critical Review. *Fibers* **2016**, *4*, 13. [CrossRef]

16. Bağcı, M. Neural Network Model for Moment-Curvature Relationship of Reinforced Concrete Sections. *Math. Comput. Appl.* **2010**, *15*, 66–78. [CrossRef]

17. Montava, I.; Irles, R.; Pomares, J.C.; Gonzales, A. Experimental Study of Steel Reinforced Concrete (SRC) Joints. *Appl. Sci.* **2019**, *9*, 1528. [CrossRef]

18. Montava, I.; Irles, R.; Segura, J.; Gadea, J.M.; Juliá, E. Numerical Simulation of Steel Reinforced Concrete (SRC) Joints. *Metals* **2019**, *9*, 131. [CrossRef]

19. Lu, Z.; Yang, Y.L.; Lu, X.; Liu, C. Preliminary study on the damping effect of a lateral damping buffer under a debris flow load. *Appl. Sci.* **2017**, *7*, 201. [CrossRef]

20. Lu, Z.; Chen, X.Y.; Zhang, D.C.; Dai, K.S. Experimental and analytical study on the performance of particle tuned mass dampers under seismic excitation. *Earthq. Eng. Struct. Dyn.* **2016**, *46*, 697–714. [CrossRef]

21. Sun, Z.; Yang, Y.; Yan, W.; Wu, G.; He, X. Moment-Curvature Behaviors of Concrete Beams Singly Reinforced by Steel-FRP Composite Bars. *Adv. Civ. Eng.* **2017**, *2017*. [CrossRef]

22. European Committee for Standardization. *EN 1994 1-1, Eurocode 4: Design of Composite Steel and Concrete Structures*; European Committee for Standardization: Brussels, Belgium, 1993.

23. European Committee for Standardization. *EN 1998, Eurocode 8: Design of Structures for Earthquake Resistance*; European Committee for Standardization: Brussels, Belgium, 2004.

Article

Retrofit Existing Frame Structures to Increase Their Economy and Sustainability in High Seismic Hazard Regions

Shuang Li [1,2,*] and Jintao Zhang [1,2]

[1]   Key Lab of Structures Dynamic Behavior and Control of the Ministry of Education, Harbin Institute of Technology, Harbin 150090, China; drzhangjintao@163.com

[2]   Key Lab of Smart Prevention and Mitigation of Civil Engineering Disasters of the Ministry of Industry and Information Technology, Harbin Institute of Technology, Harbin 150090, China

*   Correspondence: shuangli@hit.edu.cn; Tel.: +86-451-8628-2095

Received: 29 October 2019; Accepted: 10 December 2019; Published: 13 December 2019

Featured Application: The study can be used for retrofitting frame structures to satisfy a revised seismic design code, where a building site had a low seismic hazard but now has a high seismic hazard.

Abstract: The study proposes a retrofitting method with an optimum design of viscous dampers in order to improve the structural resistant capacity to earthquakes. The retrofitting method firstly uses a 2D frame model and places the viscous dampers in the structure to satisfy the performance requirements under code-specific design earthquake intensities and then performs an optimum design to increase the structural collapse-resistant capacity. The failure pattern analysis and fragility analysis show that the optimum design leads to better performance than the original frame structure. For regular structures, it is demonstrated that the optimum pattern of viscous damper placement obtained from a 2D frame model can be directly used in the retrofitting of the 3D frame model. The economic loss and repair time analyses are conducted for the retrofitted frame structure under different earthquake intensities, including the frequent earthquake, the occasional earthquake, and the rare earthquake. Although the proposed method is based on time-history analyses, it seems that the computational cost is acceptable because the 2D frame model is adopted to determine the optimum pattern of viscous damper placement; meanwhile, the owner can clearly know the economic benefits of the retrofitting under different earthquake intensities. The retrofitting also causes the frame to have reduced environmental problems (such as carbon emission) compared to the original frame in the repair process after a rare earthquake happens.

Keywords: retrofitting; viscous dampers; optimum design; collapse-resistant capacity; economic benefit; sustainability

---

1. Introduction

The devastating Mw7.9-magnitude earthquake in Wenchuan county, Sichuan Province in China on 12 May 2008 [1,2], which killed more than 87,000 people, was one of the most loss-making earthquakes in the 21st century. This event led to modifications of the design earthquake intensities in about 60 regions in the Sichuan Province in the Chinese seismic design code. Such revisions of seismic design codes often change the design earthquake intensity in a region from low to high seismic hazards. On the other hand, some types of public buildings (e.g., hospitals, schools) are required to use a higher design earthquake intensity than the code-specific intensity [3]. The reason for such changes may be due to several aspects; e.g., the new observations from recent earthquakes, technical improvements of hazard analysis, data accumulation, and societal and economic increases.

One of the issues is that a large number of existing buildings have insufficient resistant capacities to satisfy the increased design earthquake intensity. In terms of economic benefit and environmental sustainability, the best choice is to retrofit existing structures rather than demolish them and rebuild new ones. Among many retrofitting methods [4–11], the placement of viscous dampers may be one of the easy ways to improve the seismic performance of structures, because it only induces limited downtime. To date, many studies have focused on the retrofitting of structures with viscous dampers. Pekcan et al. [12] showed that the setting of the viscous damper coefficient of the story should be according to the story shear, and placements of dampers at upper stories may be not effective. Uriz and Whittaker [13] studied the effectiveness of using viscous dampers for the retrofit of a frame building before the Northridge earthquake. Silvestri and Trombetti [14] presented a parametric analysis to compare the performances offered by various systems of added viscous dampers in shear-type structures. Impollonia and Palmeri [15] studied the seismic performance of buildings retrofitted with adjacent reaction towers and viscous dampers. Karavasilis [16] studied frame structures with viscous dampers and highlighted that plastic hinges may occur in columns under earthquakes. Some studies focused on the design methods of retrofitting. Kim et al. [17] proposed a design procedure for viscous dampers based on the capacity spectrum method and verified by a 10-story and a 20-story frame structure. Lin et al. [18] presented a displacement-based seismic design method for regular buildings with viscous dampers. Habibi [19] proposed a multi-mode design method based on energy for the seismic retrofitting of structures with passive energy dissipation systems. Palermo et al. [20] proposed a "direct five-step procedure" for the seismic design of structures with viscous dampers. Zhou et al. [21] proposed a retrofitting method for frame structures with viscous dampers in order to make the structure appropriate for high-intensity earthquake environments. Although the addition of viscous dampers may induce larger residual displacements compared to other retrofitting methods (e.g., isolation, brace, etc.) in a few cases [22], the above studies demonstrate the effectiveness of structural retrofitting using dampers.

The optimum design of viscous dampers is a key issue in the retrofitting procedure based on viscous dampers. Zhang and Soong [23] presented a sequential searching procedure for the optimal placement of viscous dampers in structures. Singh and Moreschi [24] used a gradient-based method to optimally place viscous dampers in structures. In the study of Cimellaro [25], the optimal damping placement utilizing a generalized objective function is presented by accounting for displacements, absolute accelerations and base shear. Apostolakis and Dargush [26] presented a computational framework for the optimal placement of dampers in frame structures. Aydin [27] developed an optimization method to search for the optimal placement of viscous dampers based on the base moment of frame structures. Martinez et al. [28] proposed a procedure to optimally determine the damping coefficients of viscous dampers to satisfy the requirement of the maximum inter-story drift of structures. Shin and Singh [29] focused on the minimum-cost design of viscous dampers and the genetic algorithm, which was used to obtain the optimal number and placement of dampers in structures. Pollini et al. [30] also presented an effective method for seismic retrofitting using viscous dampers to achieve minimum-cost design. Parcianello et al. [31] studied a method for the optimal design of viscous dampers in order to improve the seismic behavior of structures. Lavan and Amir [32], Wang and Mahin [33] studied optimization design methods that follow the framework of performance-based earthquake engineering. Domenico and Ricciardi [34] performed an optimal design for structures with viscous dampers by a stochastic method based on the energy concept.

Viscous dampers can be employed to decrease displacements in a frame structure due to earthquakes, which allows the dissipation of seismic input energy. The design of these devices, however, is still an ongoing research topic, since it is often performed with time-history based trial-and-error methods or simplified analytical methods, which do not guarantee the optimal placements of the dampers. In addition, the existing methods have the following limitations: (1) in seismic design, it is not rational to adopt the optimum results using the minimum damping coefficient to meet the code-specific requirements under the design earthquake intensities (e.g., a frequent earthquake corresponds to an exceedance probability of 63% in 50 years, an occasional earthquake corresponds to an exceedance probability of 10% in 50 years, and a rare earthquake corresponds to an exceedance probability of 2% in 50 years), because no margin resistance is present when the structures are subjected to larger earthquake loads; (2) collapse prevention is a primary performance target for structural design [35,36]. Although the above studies may possibly provide ways to reduce the risk of structural collapse, they do not provide methods to effectively improve the collapse-resistant capacity (ground motion intensity at the structural collapse state), and thus uncertain damage may happen when the structure is under larger earthquakes; and (3) some methods derived from 2D structural models may be difficult to extend to 3D cases. Considering the above limitations, in this study, we proposed a retrofitting method with an optimum design of viscous dampers to improve the structural collapse-resistant capacity to earthquakes under the same retrofitting cost compared with ordinary retrofitting methods. The aim is to increase the structural sustainability, in an earthquake environment with a higher seismic hazard than in the initial structural design.

2. Retrofitting Method Using Viscous Dampers

2.1. Structural Model Used for Verification

In order to verify the retrofitting method, a 6-story building is used with regular configurations in plan and elevation. This frame structure was designed according to the CSDB code [3] with the design earthquake environment of Region 6 (Peak Ground Acceleration, i.e., PGA (Peak Ground Acceleration) = 0.05 g, corresponding to a 10% exceedance probability in 50 years, with a type II site condition), and now it should meet the requirements of the design earthquake environment of Region 8 (PGA = 0.2 g corresponding to a 10% exceedance probability in 50 years, with a type II site condition) due to design code revision. The geometry layout of the frame structure is shown in Figure 1. More information on the reinforced concrete (RC) frame, such as cross-section dimensions of columns and beams and usage of steel rebars, is provided in Tables 1 and 2. The frame is a flexible structure with fundamental and second periods $T_1 = 1.88$ s and $T_2 = 0.62$ s, respectively.

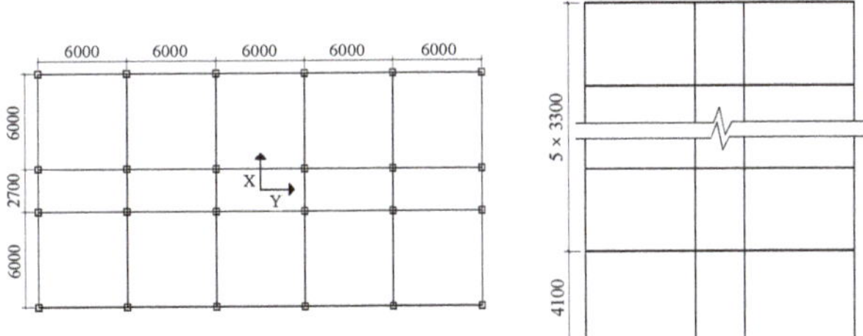

Figure 1. Layout of frame structure (unit: mm).

Table 1. Dimensions of beams and rebars (X direction).

Story	Size (mm × mm) Width × Height		Area of Longitudinal Rebars (mm^2)/Stirrup			
	Side Bay	Middle Bay	Side Bay		Middle Bay	
			Beam Ends	Midspan	Beam Ends	Midspan
1	200 × 50	200 × 300	2250/ϕ8@100	1600/ϕ8@200	2250/ϕ8@100	1350/ϕ8@200
2	200 × 50	200 × 300	2200/ϕ8@100	1600/ϕ8@200	2200/ϕ8@100	1250/ϕ8@200
3	200 × 50	200 × 300	2050/ϕ8@100	1600/ϕ8@200	2050/ϕ8@100	1100/ϕ8@200
4	200 × 50	200 × 300	1800/ϕ8@100	1650/ϕ8@200	1800/ϕ8@100	850/ϕ8@200
5	200 × 50	200 × 300	1450/ϕ8@100	1650/ϕ8@200	1450/ϕ8@100	750/ϕ8@200
6	200 × 50	200 × 300	850/ϕ8@100	1400/ϕ8@200	850/ϕ8@100	600/ϕ8@200

Note: Beams in Y direction have the same usage of rebars as that in X direction.

Table 2. Dimensions of columns and rebars.

Story	Size (mm × mm)		Area of Longitudinal Rebars (mm²)/Stirrup	
	Side Column	Middle Column	Side Column	Middle Column
1	400 × 400	400 × 400	2000/ϕ8@100	1400/ϕ8@100
2	400 × 400	400 × 400	800/ϕ8@100	1000/ϕ8@100
3	400 × 400	400 × 400	800/ϕ8@100	1000/ϕ8@100
4	400 × 400	400 × 400	700/ϕ8@100	700/ϕ8@100
5	400 × 400	400 × 400	700/ϕ8@100	700/ϕ8@100
6	400 × 400	400 × 400	1300/ϕ8@100	1000/ϕ8@100

The seismic response analysis is conducted using the OpenSees program [37]. The nonlinearBeamColumn elements with fiber section division are used to model beams and columns, while Concrete02 (Kent-Scott-Park) and Steel02 (Menegotto and Pinto) material models are used for the concrete and rebars [38]. The compression and tension strengths and strain at the peak stress of concrete are 20.1 Mpa, 2.01 Mpa and 0.002, respectively. The yielding strengths and elastic modulus of the longitudinal rebars are 335 Mpa and 2.0×10^5 Mpa. P-Δ coordinate transformation is used to consider the geometric nonlinearity. The shear failure is not considered in the element model. This assumption will not introduce large errors, because the columns and beams are required to be designed such that flexure damage occurs before shear damage. The damping ratio is set as 0.05 in the analyses.

The force provided by a viscous damper is calculated, in general, as

$$F = Cv^\alpha \tag{1}$$

where C is the damping coefficient, α is the velocity index, and v is the relative velocity. In the following numerical example, α is set as 1.0 corresponding to a linear viscous damper. The TwoNodeLink element with Maxwell material in OpenSees is used to model the viscous dampers [37]. The Maxwell material model comprises a series-wound stiffness spring and damping pot. The force from the damping pot is calculated by using Equation (1), and the stiffness is commonly selected within 2~3 times the value of coefficient (2.5 times is used in this study). Based on the survey of some companies' products, the commonly used damping coefficients of viscous dampers are 20, 50, 100, 150, 200, 250, 300, 350, 400, 500, 600, 750, 800, 1000, 1200, 1300, 1500, 1600, 1800, 2000, 2500, 2600, 2800, 3000, 3300, 3500, 3600, 3800, 4000, 4800, 5000, 5200, 5600, 6000, 6700, 8000, 8700, 9300, 10,000, 12,000, 15,000, 18,000, 20,000, 24,000, 26,000, 28,000, and 30,000 kN·(s/m). These parameters can be referred in the retrofitting procedure.

Ground motions (GMs) are selected from the PEER (Pacific Earthquake Engineering Research Center) NGA (Next Generation Attenuation) strong motion database (http://ngawest2.berkeley.edu) to be compatible with the design spectrum in the CSDB code [39] at the main period points. Since a suite of 3 to 11 GMs is usually recommended in codes [39–41] to get an accurate assessment of structural response in seismic design procedures, a total of 7 GMs (see Table 3) are used in this study. The matching to the design spectrum of the building site is shown in Figure 2.

Table 3. Selected ground motions (GMs).

Tag	Earthquake	NGA ID and Component	Year	Magnitude
1	San Fernando, USA	RSN68 SFERN PEL090	1971	6.61
2	San Fernando, USA	RSN93 SFERN WND143	1971	6.61
3	Tabas, Iran, USA	RSN143 TABAS TAB-T1	1978	7.35
4	Imperial valley—06, USA	RSN161 IMPVALL.H H-BRA315	1979	6.53
5	Imperial valley—06, USA	RSN178 IMPVALL.H H-E03230	1979	6.53
6	Imperial valley—06, USA	RSN180 IMPVALL.H H-E05140	1979	6.53
7	Imperial valley—06, USA	RSN183 IMPVALL.H H-E08140	1979	6.53

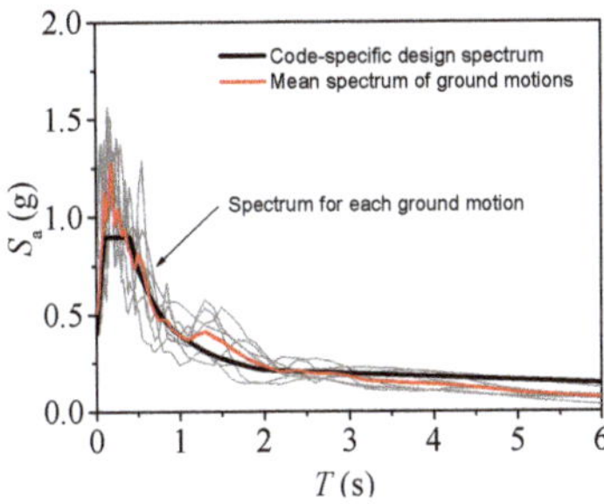

Figure 2. Selected GMs compatible with the design spectrum.

2.2. Design Method

A 2D model (a planar frame in the X direction) is firstly used in this section to illustrate the retrofitting method and while a 3D model will be adopted in Section 3. Figure 3 shows a typical placement layout of viscous dampers in the structure. The retrofitting method can be conducted through three steps (Step 1: Initial placements of dampers to satisfy code-specific requirements; Step 2: Optimum placements of the dampers under same retrofitting cost; and Step 3: Final check based on code-specific requirements). These steps will be discussed in detail in the following.

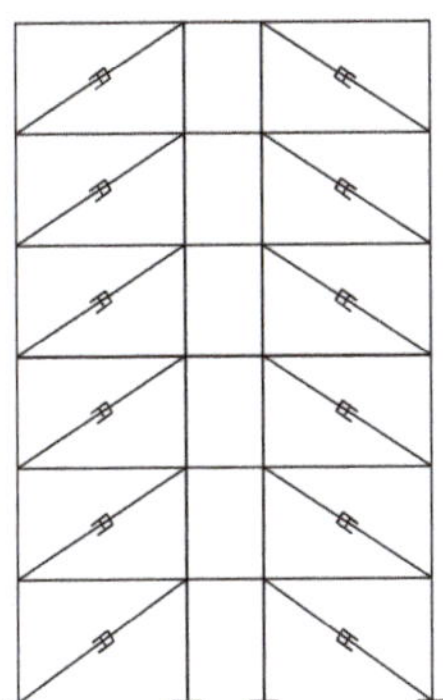

Figure 3. Placement layout of viscous dampers in the structure (planar frame for illustration).

2.2.1. Step1: Initial Placements of Dampers to Satisfy Code-Specific Requirements

Code-specific requirements under rare earthquakes are used in the initial design of dampers. The inter-story drift ratio of story I (δ_i) is used as the requirement, by considering that δ_i should be less than 1/50 for RC frame under rare earthquakes [3]. In Region 8, design intensities are PGA = 0.07 g, 0.2 g, and 0.4 g for frequent, occasional and rare earthquakes, respectively. Note that Step 1 can be implemented with any procedure provided that the design requirements are met. The following procedure is suggested:

(1) For each GM, scale PGA to 0.4 g. Select a layout pattern of the damper placement; e.g., the pattern shown in Figure 3.

(2) Use a uniform placement from a small damping coefficient, e.g., $C_i = 20$ kN·(s/m), as shown in Section 2.1 or from any other value based on the designer's experience.

(3) Calculate δ_i for each story. If $\delta_i \leq 1/50$ for any story, the initial placement of dampers satisfies code-specific requirements; stop the procedure, otherwise, increase the damping coefficients C_i until the requirement $\delta_i \leq 1/50$ is met for story i. This procedure is performed until $\delta_i \leq 1/50$ for any story. In such a way, C_i of each story is determined.

The use of the above procedure results in 7 schemes for initial placement patterns of dampers corresponding to the 7 GMs listed in Table 3. Table 4 provides the initial placements of dampers to satisfy code-specific requirements. Large discrepancies are observed in both the distribution pattern of damping coefficients among stories and the total damping coefficient (sum of damping coefficients of all stories). This observation reveals that although the GMs are compatible with the design spectrum, different demands may arise from the uncertainties of GMs. For example, to satisfy code-specific requirements, a total damping coefficient of 4,170,000 N·(s/m) is needed for GM 3 while only 440,000 N·(s/m) is needed for GM 2.

Table 4. Initial placements of dampers to satisfy code-specific requirements (unit: N·(s/m)).

Retrofitting Scheme	C_1	C_2	C_3	C_4	C_5	C_6	$\sum_{i=1}^{6} C_i$
Initial scheme 1	300,000	20,000	250,000	250,000	20,000	20,000	860,000
Initial scheme 2	150,000	150,000	50,000	50,000	20,000	20,000	440,000
Initial scheme 3	1,300,000	1,300,000	1,000,000	500,000	50,000	20,000	4,170,000
Initial scheme 4	1,000,000	1,000,000	800,000	250,000	20,000	20,000	3,090,000
Initial scheme 5	1,200,000	1,300,000	750,000	200,000	50,000	50,000	3,550,000
Initial scheme 6	800,000	750,000	600,000	100,000	20,000	20,000	2,290,000
Initial scheme 7	20,000	20,000	500,000	50,000	20,000	20,000	630,000

The incremental dynamic analysis (IDA) [42] is performed to check the collapse-resistant capacity of the frame after the initial placements of dampers. The structural collapse state is determined according to FEMA (Federal Emergency Management Agency) 350 [43], i.e., the 20% tangent slope method, which defines the point on the IDA curve with a tangent slope that reduces to 20% of the initial elastic slope as the collapse capacity point, or according to the inter-story drift ratio (IDR) is larger than 10%. The collapse-resistant capacities are $S_a(T_1)$ = 0.16 g, 0.14 g, 0.22 g, 0.28 g, 0.19 g, 0.19 g, and 0.22 g after retrofitting by the 7 schemes, with increases of only 0.01 g, 0.0 g, 0.08 g, 0.03 g, 0.07 g, 0.05 g, and 0.0 g with respect to the original frame.

2.2.2. Step 2: Optimum Placements of the Dampers at the Same Retrofitting Cost

The initial design of dampers will be further optimized in this section to increase the structural collapse-resistant capacity at the same retrofitting cost. It is assumed that the cost of retrofit is proportional to the total damping coefficient [14,34]. Therefore, in the optimization, the total damping coefficient obtained in Step 1 is kept unchanged. The following procedure can be followed:

(1) By referring to initial scheme 1, perform IDA analysis on the structure to obtain its collapse-resistant capacity CC. Note that the selected GM for initial scheme 1 corresponds to GM 1.
(2) Scale GM to CC and perform a further dynamic analysis on the initial retrofitted structure. Obtain the inter-story drift ratio (δ_i) of each story.
(3) If $\delta_{i,\,max}$ is close enough to the pre-defined collapse inter-story drift ratio $\delta_{collapse}$, no iteration is necessary; otherwise, modify the distribution of the story damping coefficient C_i. The story damping coefficient C_i is reduced in stories with δ_i lower than $\delta_{collapse}$ and increased in stories with δ_i higher than $\delta_{collapse}$ (using Equation (2) for the adjustment and Equation (3) to keep the total damping coefficient unchanged). The adjustment is repeated until all the inter-story drift ratios satisfy Equation (4).

$$C_i^{j+1} = \left(\frac{\delta_i^j}{\delta_{average}}\right)^{\alpha} \cdot C_i^j \tag{2}$$

$$\sum_{i=1}^{6} C_i = C_{total} \tag{3}$$

$$\delta_i^{j+1} < \delta_{collapse} \tag{4}$$

In the above equations, C_i^{j+1} and C_i^j are the damping coefficients of dampers placed at story I after j+1th and jth adjustments; δ_i^{j+1} and δ_i^j are the inter-story drift ratios at story I after j+1th and jth adjustments under the scaled GM; $\delta_{average}$ is the average value of inter-story drift ratios of all stories; and α is a coefficient between 0 and 1 that controls the convergence gradient. In the study, α is set to 0.3; and C_{total} is the total damping coefficient of the initial retrofitted structure. Please note that 6, appearing in Equation (3), denotes the number of stories of the example frame. $\delta_{collapse}$ is the inter-story drift ratio to determine the collapse state of the optimum structure. According to FEMA 356 [44] and to experimental studies [45–47], stating that the inter-story drift ratio corresponding to "Collapse Prevention" state is in the range 2–4%, $\delta_{collapse}$ is set to 4% in this study. Note that 10% is used previously in the IDA analyses to check a real collapse state, whereas 4% is used here to assess a "Collapse Prevention" state which is actually a nominal collapse suitable to practical design.

(4) Perform IDA analysis on the structure to obtain its new collapse-resistant capacity CC'. Scale the PGA of GM to CC' and perform (3) again. Carry out this procedure until any further iteration is unable to increase the CC'. The value of CC' so obtained is recognized as the optimum result for the selected GM.

(5) Using the above procedure, 7 new schemes of placements of dampers for the frame can be obtained; these new schemes have the same retrofitting cost as the initial retrofitting schemes obtained in Step 1, but with higher collapse-resistant capacities.

For brevity, only the new scheme 1 is provided here for illustration. Table 5 shows the changes in the damping coefficients of each story in Step 2. It is observed that the damping coefficients of lower stories become larger while those of upper stories become smaller, and that there are no requirements for placements of dampers at the top two stories. This finding is consistent with the study of Pekcan et al. [12]. The collapse-resistant capacity increases to 0.03 g compared to the initial retrofitted frame. Figure 4 shows the change of the inter-story drift ratio of each story, evidencing how the collapse-resistant capacity increases in the procedure. It is shown that for the initial retrofitted schemes, one or two stories of the structure collapse under the GMs. With the optimum design, more than one story collapses nearly at the same ground motion intensity in contrast with what occurred for the initial retrofitted frame, in which a story collapses while other stories still remain in the elastic range. Figure 5 shows that, compared with the initial retrofitted schemes obtained in Step 1, the procedure carried out in Step 2 leads to an increase of the collapse-resistant capacities of the 7 schemes of 18.75, 7.14, 13.64, 28.57, 5.26, 47.4, and 13.64%, respectively. Note that the IDA curves shown in Figure 4 are presented only to illustrate the procedure proposed in the study. In the practical design, for saving the computational cost, the IDA analyses can start from a spectral value S_a corresponding to PGA = 0.4 g because the structural collapse-resistant capacity is surely larger than that intensity.

Table 5. Variation of damping coefficients in the optimum procedure (using Initial scheme 1 as an example) (unit: N·(s/m)).

Optimum Times	Collapse Resistance	C_1	C_2	C_3	C_4	C_5	C_6	$\sum_{i=1}^{6} C_i$
Initial scheme 1	0.16 g	300,000	20,000	250,000	250,000	20,000	20,000	860,000
1st time	0.18 g	499,230	32,883	185,666	127,186	7857	7178	860,000
2nd time	0.19 g	849,740	1476	5545	3329	0	0	860,000
Optimum scheme 1	0.19 g	849,740	1476	5455	3329	0	0	860,000

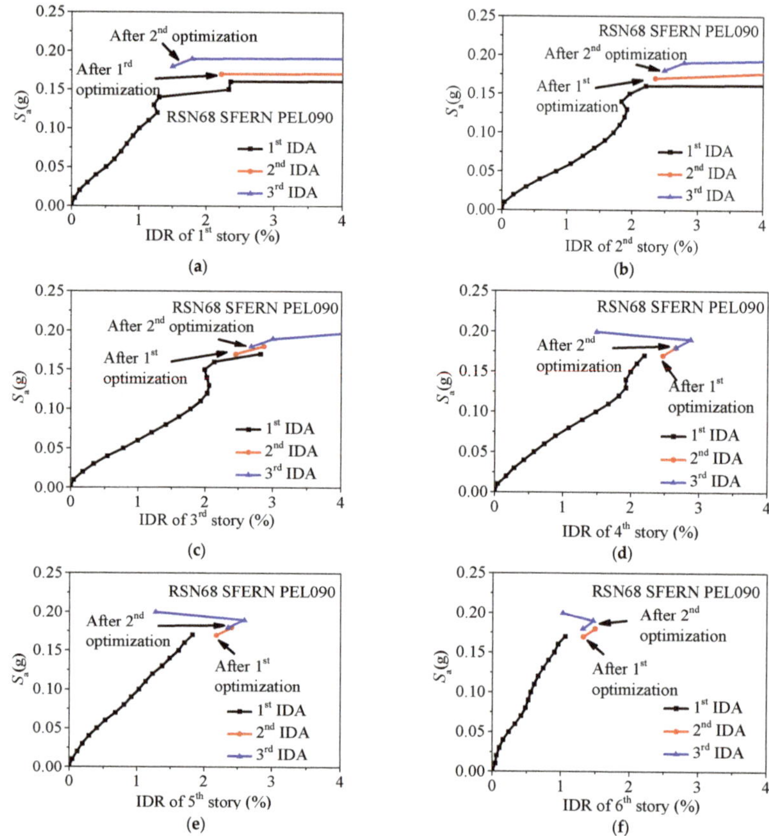

Figure 4. Variation of the collapse resistance in the optimum procedure (using Initial scheme 1 as an example; IDR: Inter-story drift ratio): (**a**) 1st story, IDA (incremental dynamic analysis) curves; (**b**) 2nd story, IDA curves; (**c**) 3rd story, IDA curves; (**d**) 4th story, IDA curves; (**e**) 5th story, IDA curves; (**f**) 6th story, IDA curves.

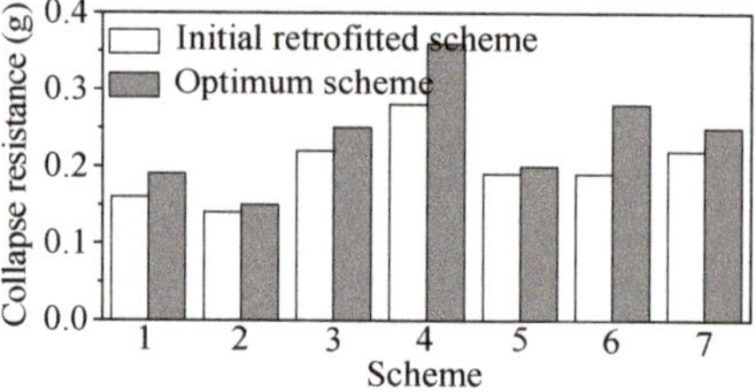

Figure 5. Comparison between the structural collapse resistance of the initial retrofitted scheme and the optimum retrofitted scheme.

2.2.3. Step 3: Final Check Based on Code-Specific Requirements

In the above two steps, the results of a retrofitted scheme under a given GM have been provided. In this section, a total of 7 GMs will be used together to get the final optimum retrofitted scheme. At first, the minimum total damping coefficient is obtained; then, the optimum distribution of this total damping coefficient along the height of the frame is determined. The following procedure is performed:

(1) Determine the required minimum total damping coefficient. Note that the schemes obtained in Step 2, which satisfy the code-specific requirements about inter-story drift ratio under the rare earthquake intensity, are taken as schemes to be considered for each GM. In this step, each scheme obtained in Step 2 is analyzed under 7 GMs scaled to PGA = 0.07 g for frequent earthquake and PGA = 0.4 g for rare earthquake, respectively. Then, the average inter-story drift ratio of each scheme is obtained by averaging 7 results. The required minimum total damping coefficient is determined from the scheme average inter-story drift ratios, which satisfies the code requirements under frequent and rare earthquakes (1/550 and 1/50, respectively [3]).

(2) Determine an optimum distribution pattern from schemes obtained in Step 2 with total damping coefficient scaled to the required minimum total damping coefficient. This procedure checks which distribution pattern is better under a given minimum total damping coefficient. The verification criteria are the same used in (1).

Using the above procedure, the final optimum scheme will be obtained. For the example frame, the required minimum total damping coefficient is 3,550,000 N·(s/m). Scaling all the other schemes' total damping coefficients to this value, and checking the average inter-story drift ratios again, the final optimum retrofitted scheme is obtained and referred to as scheme 5, as shown in Table 6. Figure 6 shows that the optimum retrofitted scheme satisfies the code-specific requirements for the average inter-story drift ratio under frequent and rare earthquake intensities (1/550 and 1/50, respectively [3]).

Table 6. Damping coefficients in the optimum retrofitted scheme (unit: N·(s/m)).

Strengthen Scheme	C_1	C_2	C_3	C_4	C_5	C_6	$\sum_{i=1}^{6} C_i$
Optimum scheme 1	849,740	1476	5455	3329	0	0	860,000
Optimum scheme 2	262,542	115,899	31,765	20,796	5172	3826	440,000
Optimum scheme 3	3,335,742	543,619	164,958	125,681	0	0	4,170,000
Optimum scheme 4	2,741,236	186,395	101,966	60,403	0	0	3,090,000
Optimum scheme 5	1,610,247	1,339,753	553,541	46,459	0	0	3,550,000
Optimum scheme 6	1,604,274	412,497	139,411	133,818	0	0	2,290,000
Optimum scheme 7	602,094	3659	22,695	1551	0	0	630,000

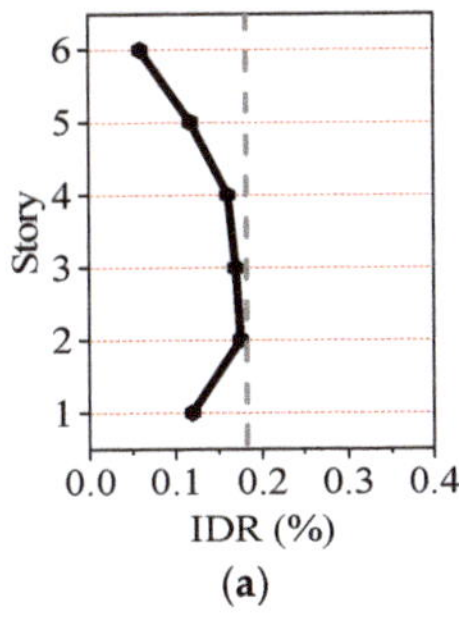

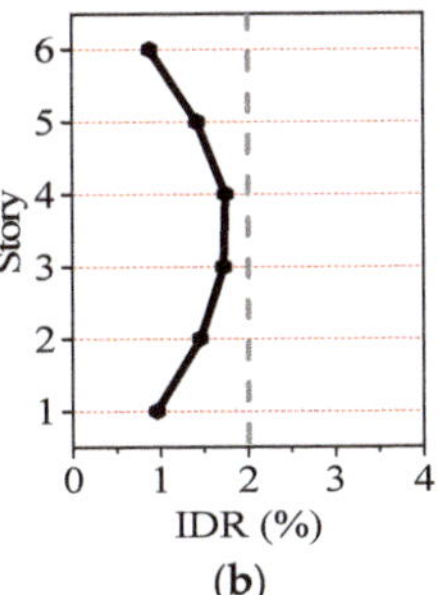

Figure 6. Average inter-story drift ratio (IDR) under frequent and rare earthquakes: (**a**) Frequent earthquake; (**b**) Rare earthquake.

2.3. Comparison of Failure Pattern and Collapse Fragility

This section verifies the seismic performance of the frame after retrofitting with the optimum scheme. The yielding occurs if the cross-section curvature response is larger than the yielding curvature value. Note that the concrete and rebar in a cross-section may be the same, but different axial compression ratios in columns among different stories lead to the cross-section curvatures not being the same in different stories. The calculated yielding curvatures for the side columns from the 1st to

6th story are 0.0111, 0.0099, 0.0090, 0.0092, 0.0078, and 0.0067 m^{-1}, and those for the center columns are 0.0107, 0.0113, 0.0102, 0.0102, 0.0085, and 0.0069 m^{-1}.

Figures 7 and 8 show the maximum curvature ductility μ_{max} (maximum value/yielding value) of columns in each story. The hollow red circle shows that the yielding is not achieved where $\mu_{max} < 1$. The solid red circle shows that the yielding is achieved where $\mu_{max} \geq 1$. In order to provide a clear demonstration, the viscous dampers placed on the structure are not displayed in the figures. The figures show that, for the initial retrofitted structure, the yielding occurs and the collapse occurs at the bottom section of the firs story columns under anyone of the 7 GMs, while the upper stories remain elastic. As a contrast, for the optimum retrofitted structure, yielding does not occur under GM 1, GM 2, and GM 7. Meanwhile, it seems that the yielding moves from the bottom story to upper stories. Compared to the initial retrofitted structure, the yielding occurs more uniformly, and the maximum curvature ductility μ_{max} decreases by 51.6%~77.7% at the same locations. Therefore, the optimum retrofitted structure has a smaller probability of collapse. This figure clearly illustrates the effectiveness of the optimum retrofitted procedure.

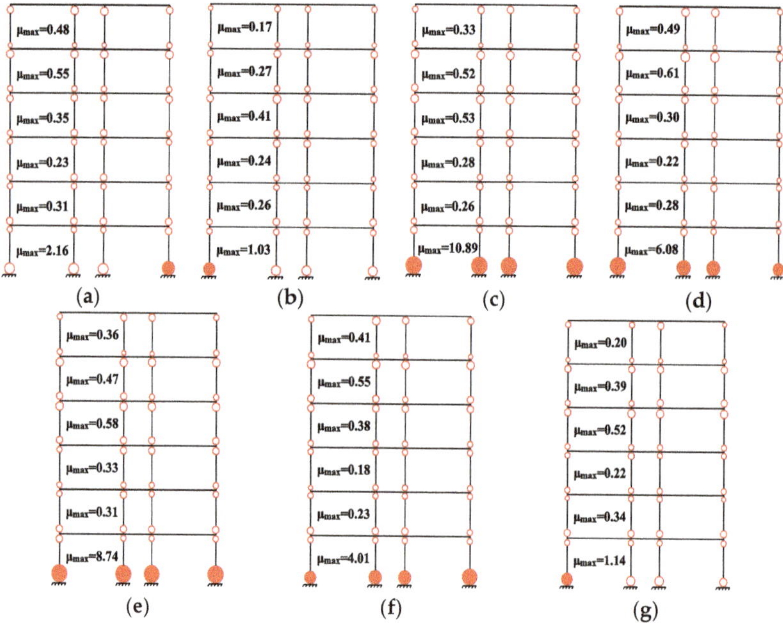

Figure 7. Locations of yielding in columns of initial retrofitted frame (Initial scheme 5): (**a**) GM 1; (**b**) GM 2; (**c**) GM 3; (**d**) GM 4; (**e**) GM 5; (**f**) GM 6; (**g**) GM 7.

The collapse fragility, which is calculated using Equation (5) [48], is adopted to compare the change of the collapse-resistant capacity of the frame.

$$P(\text{Collapse}|\text{IM} = im) = \Phi\left(\frac{\ln(x/m_R)}{\beta_R}\right) \tag{5}$$

Here is the standard normal distribution function; $P(\text{Collapse}|\text{IM} = im)$ is the probability of collapse under GM with intensity im; and m_R and β_R are the median value and logarithmic standard deviation of the fragility. The collapse frequency shown by Equation (6) can be also used to calculate the fragility; then, n_c and N_{total} are calculated by statistical data, and finally fitted by Equation (5) to obtain a smooth fragility curve.

$$P(\text{Collapse}|\text{IM} = im) = \frac{n_c}{N_{total}} \tag{6}$$

In Equation (6), n_c is the number of collapse cases under GM with *im* (intensity measure); N_{total} is the number of total dynamic analyses under GM with *im*. The acceleration spectral intensity $S_a(T_1, \xi = 5\%)$ is used as *im* in the analyses.

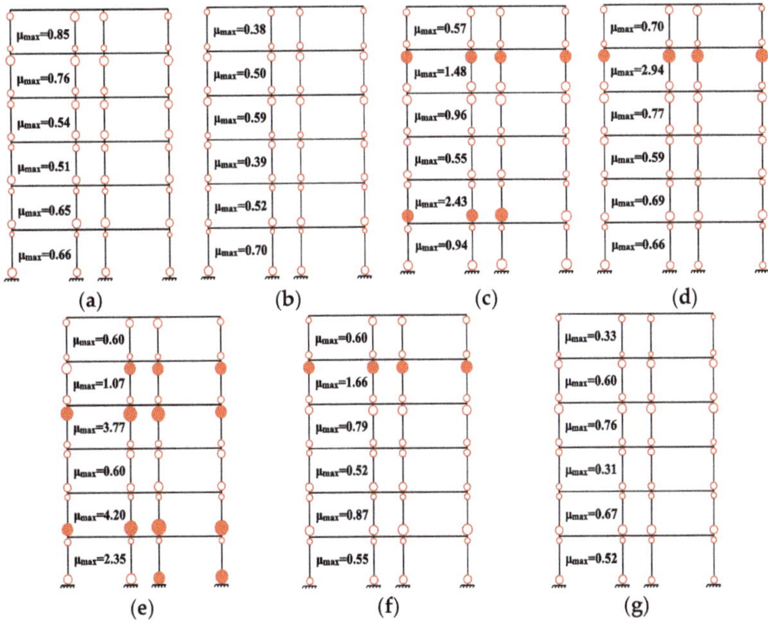

Figure 8. Locations of yielding in columns of optimum retrofitted frame: (a) GM 1; (b) GM 2; (c) GM 3; (d) GM 4; (e) GM 5; (f) GM 6; (g) GM 7.

Figure 9 shows the fragility curves of the original frame, initial retrofitted frame, and the optimum retrofitted frame. In this figure, the result of the original frame was also provided because we want to illustrate both the retrofitting effect on the original structure and the optimum effect on the initial retrofitted frame. The figure shows that the collapse probability of the optimum retrofitted frame is the smallest, which demonstrates that the placements of viscous dampers can increase the structural collapse-resistant capacity. The maximum decline of collapse probability between the original and optimum retrofitted frames is 77.26%, and that between the initial retrofitted and optimum retrofitted frames is 33.53%. Note that the retrofitting costs of the initial retrofitting and optimum retrofitting are the same.

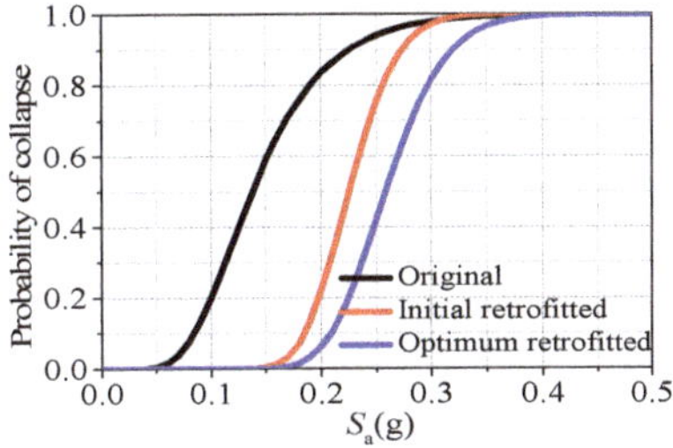

Figure 9. Collapse fragility curves of the 2D frame.

3. Extension to 3D Structures

The practical application of retrofitting is always on an actual 3D structure. This section will provide a way to extend the 2D results to 3D structures, which will significantly reduce the computational cost in the optimum retrofitting procedure. The 3D frame shown in Figure 1 is with 3-bay in X direction and 5-bay in Y direction. The dampers will only be installed at a few places in a 3D structure in order to minimally disturb the structural functions. Figure 10 shows the bays at which viscous dampers are placed in X direction and Y direction.

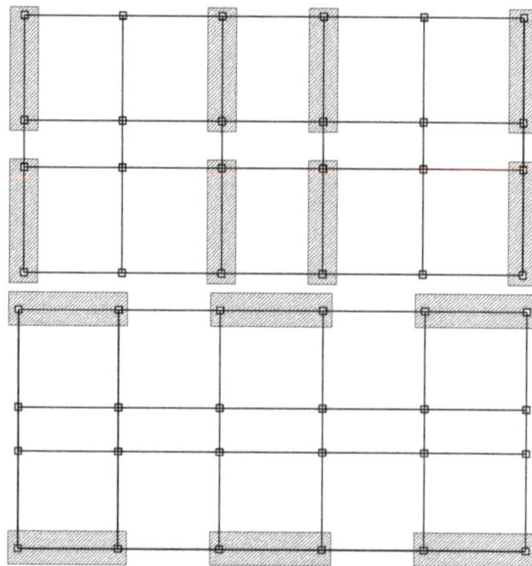

Figure 10. Placement of viscous dampers in X and Y directions (placed at shadow bays).

Since the distributions of the story lateral stiffness and story strength of the 3D frame are similar to those of the planar frame used in Section 2, the distribution patterns of additional story damping coefficients can be the same as those in the 2D case, and the values can be proportional to the results obtained by using the 2D case. As stated in Section 2.2.3, scheme 5 in Table 6 is the optimum scheme in the 2D case. Table 7 provides the damping coefficients used in the retrofitting in X direction and Y direction, respectively. Note that these values are for one of the planar frames in the 3D frame; i.e., 1/4 of the total damping coefficient used in X direction and 1/2 of the total damping coefficient used in Y direction. The damping coefficients in X direction are calculated as 1.5 times the values obtained in the 2D case. For the damping coefficients in Y direction, the distribution pattern is fixed but the values need to be determined. By using a trial-and-error method, the total damping coefficient is determined to be 8,875,000 N·(s/m). The ground motions are applied in both X direction and Y direction. Figure 11 shows the average inter-story drift ratios under the frequent earthquake and the rare earthquake in X direction and Y direction, which indicate that the structural performance satisfies the code-specific requirements (1/550 under frequent earthquake and 1/50 under rare earthquake). Collapse fragility analyses are also conducted on the initial retrofitted frame and optimum retrofitted frame in X direction and Y direction which reveal that, with the same retrofitting cost, the optimum retrofitted frame has a larger collapse-resistant capacity with about 15% increases compared with the initial retrofitted frame.

Table 7. Optimum damping coefficients of 3D frame (unit: N·(s/m)).

3D Frame	C_1	C_2	C_3	C_4	C_5	C_6	$\sum_{i=1}^{6} C_i$
X direction	2,415,000	2,010,000	825,000	75,000	0	0	5,325,000
Y direction	4,025,000	3,350,000	1,375,000	125,000	0	0	8,875,000

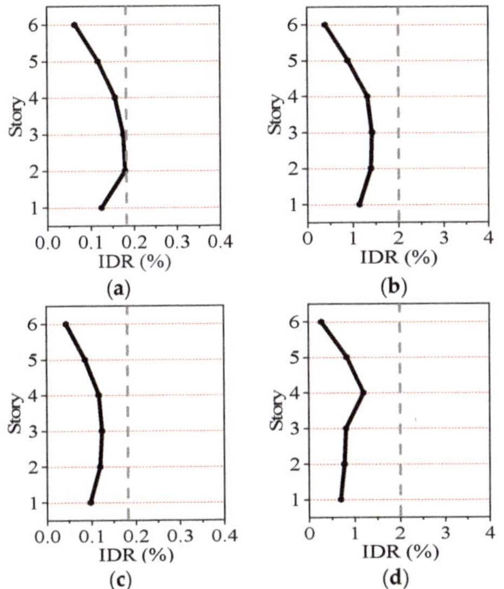

Figure 11. Average inter-story drift ratio (IDR) under frequent and rare earthquakes: (**a**) *X* direction, frequent earthquake; (**b**) *X* direction, rare earthquake; (**c**) *Y* direction, frequent earthquake; (**d**) *Y* direction, rare earthquake.

4. Economic Analysis for the Proposed Method

The above study focuses on the seismic performance of the retrofitted frame. The economic benefit should be determined, as it has an impact on whether the retrofitting can be conducted or not. Therefore, this section discusses the economic benefit of the retrofitting of existing frame structures from low to high seismic hazard regions. The methodology proposed in FEMA P-58 and the PACT software [49] are used to conduct the economic analysis. Note that we have not investigated any other retrofitting solution, such as seismic isolation, and so the economic benefit analysis is only assessed for retrofitting with viscous dampers.

4.1. Performance Group of Structural and Non-Structural Members

The replacement cost is calculated as $(249.03 \times 441 \times 6)/10^4 = 65.9$ MUD (MUD represents one million/100 US dollars), in which 249.03 US dollars is the statistical construction cost of 1 m^2 at the building location [50], 441 m^2 is the slab area of one story (see Figure 1), and 6 is the number of stories. The total replacement cost includes the core, shell, and all tenant improvements and contents [49]. Considering that the building's function is a middle school, the total replacement cost is finally determined to be 76.7 MUD, including the cost due to structural replacement 65.9 MUD, and costs due to the damage of electronic teaching platforms, projectors, tables and chairs, blackboards, and computers (numbers and costs are provided in Table 8). Regarding the replacement time, including the foundation, upper structure, decoration, and elevator installation, etc., Table 9 provides the required workdays for each task according to the quota of construction information [51].

Table 8. Performance groups considered in the loss estimation of 6-story RC (reinforced concrete) frame.

Group	Fragility ID [49]	EDP	Cost/Each (Unit: US Dollar)	Number in Each Story	Total Number
Teaching platform	E2202.020	PFA	194.3	10	60
Projector	C3033.002	PFA	448.4	10	60
Table and chair	E2022.020	PFA	14.9	300	1800
Blackboard	E2022.021	PFA	29.9	10	60
Computer	E2022.022	PFA	448.4	10	60

Table 9. Construction periods of key items of the 6-story RC frame.

Construction Items	Construction Period (Unit: Workday)
Foundation project	30
Upper structure project	305
Decorate project	60
Elevator system installation	50
Heat-supply system installation	50
Air conditioning system installation	65
Electric power substation project	40
Total required workdays	600

In the PACT software [49], members cataloged in the same performance group correspond to the same fragility curve. Table 10 provides the statistical structural and non-structural member information of the building. For the structural members, only the beam-column joint is set as a performance group because damage commonly occurred at the beam and column ends of the frame structures. Table 10 also gives the EDPs (Engineering Demand Parameters) used in the evaluation. IDR means that the member damage is sensitive to the inter-story drift ratio, and PFA means the member damage is sensitive to the peak floor acceleration. The fragility ID shows the fragility curves selected from the PACT software corresponding to the performance group. The costs of the performance groups in Table 10 are determined by using the built-in data in PACT software. Note that the performance groups in Table 8 are not built-in data in the PACT software, only similar fragility IDs are used and the costs are determined by their market prices.

Table 10. Statistical data on the performance groups of the 6-story frame.

	Performance Group	Fragility ID [49]	EDP	Costing Based Upon in PACT [49]	Number in Each Story	Total Number
Structural member	Beam joints	B1041.101b	IDR	1 EA	24	144
	Curtain walls	B2022.001	IDR	30 SF	34.82	208.92
	Partitions	C1011.001a	IDR	100 LF	5.32	31.92
	Wall finishes	C3011.001a	IDR	100 LF	1.34	8.04
	Cold water piping	D2021.011a	PFA	1000 LF	0.07	0.42
	Hot water piping	D2022.011a	PFA	1000 LF	0.40	2.40
		D2022.021a	PFA	1000 LF	0.16	0.96
	Sanitary waste piping	D2031.011b	PFA	1000 LF	0.21	1.26
	HVAC equipment	D3041.011a	PFA	1000 LF	0.24	1.44
Non-structural member		D3041.031a	PFA	10 EA	2.37	14.22
	Variable Air Volume (VAV) box	D3041.041a	PFA	10 EA	1.90	11.4
	Concrete tile roof	B3011.011	PFA	100 SF	32.28	32.28
	Recessed ceiling lighting	C3033.001	PFA	1 EA	71.21	427.26
	Independent pendant lighting	C3034.001	PFA	1 EA	71.21	427.26
	Sprinkler water supply	D4011.021a	PFA	1000 LF	0.85	5.1
		D4011.031a	PFA	100 EA	0.38	2.28
	Stairs	C2011.001a	IDR	1 EA	2.00	12
	Low voltage switchgear	D5012.021a	PFA	225 AMP	1.00	6
	Elevator	D1014.011	PFA	1 EA	1.00	1.00
	Motor control center	D5012.013a	PFA	1 EA	2.00	2.00

Note: HVAC, Heating Ventilation Air Conditioning.

The cost of viscous dampers is proportional to the damping coefficient. After inquiring with several damper companies, the costs for viscous dampers in a single bay along the structural height in X direction were determined to be 4085, 3400, 1395, 127, 0, and 0 (a total of 9006 US dollars), and the costs for viscous dampers in a single bay along the structural height in Y direction are 6807, 5666, 2326, 211, 0, and 0 (a total of 15010 US dollars). Therefore, the total cost of all viscous dampers is (9006 × 2

$\times\, 4 + 15010 \times 3 \times 2)/10^5 = 16.2$ MUD. The installation of a viscous damper needs 2 workers with 1 workday, and the cost is estimated as $45 \times 2 = 90$ US dollars, in which 45 US dollars are estimated as the salary for 1 worker per day at the building's location. The estimated total number of required viscous dampers is 56, hence the total installation cost is $(90 \times 56)/10^4 = 0.5$ MUD. Therefore, the total retrofitting cost of the 3D structure is $16.2 + 0.5 = 16.7$ MUD.

4.2. Economic Benefit Analysis

Since the structure is retrofitted considering an increase of seismic hazard, according to Chinese seismic codes, from Region 6 to Region 8 [3], the economic benefit is calculated for both the original and the retrofitted frames under the earthquake intensity corresponding to Region 8 (PGA = 0.07 g, 0.2 g, and 0.4 g for the frequent earthquake, the occasional earthquake and the rare earthquake). Note that the results of a large number of analyses would provide a smoothed distribution for the probabilistic evaluation of earthquake results. However, with the current state of modeling capability, such a method would be infeasible for implementation in practice. Instead, the PACT software uses the Monte Carlo procedure to evaluate possible outcomes given a limited set of inputs [49]. In this procedure, limited suites of analyses are performed by using actual GMs to derive a statistical distribution of demands using many of building response states for a specific intensity of motion. In the present study, 1000 Monte Carlo realizations were conducted for a GM intensity level. The GMs selected in Section 2.1 are used here.

Figure 12a–f compares the repair costs between the original frame and retrofitted frame. These figures show that the probabilities are smaller than the specific repair cost levels. Figure 13 shows the median repair costs, which are the values corresponding to a probability of 50% in Figure 12. In the figure, we added the retrofitting cost (16.7 MUD) into the total repair cost in order to compare the economic benefit. The figure also compares the repair costs of structural members and non-structural members, respectively. According to FEMA P-58 [49], when the structure is seriously damaged, as in the case shown in Figure 13c, it is necessary to demolish the structure and remove it from the site. Hence, the repair cost (or the replacement cost, which is more rational here) includes the cost to demolish the damaged structure and clear the debris in addition to replacing the structure "in-kind". Demolition and site clearance will increase structural repair costs by up to 20–30%. Under the rare earthquake excitation, the original frame collapses; therefore, the total repair cost increases by 20% in the calculation, i.e., $76.7 \times 120\% = 92.04$ MUD. Hence, the ratio between the cost of complete replacement and the cost of retrofitting is about 45%.

The following observations can be made from these figures. (1) Under a frequent earthquake, for both the original and retrofitted frames, the losses are due to non-structural members, while the structural members' damages are negligible. (2) Under an occasional earthquake, losses are due to both the structural and non-structural members in the original frame, in which the loss due to non-structural members makes up a large proportion. The retrofitted frame suffers no loss due to the structural members. If the initial retrofitting cost is not considered, the loss of the retrofitted frame is smaller than that of the original frame. (3) Under a rare earthquake, for both the original and retrofitted frames, the losses are due to both the structural and non-structural members. The repair cost due to structural members accounts for a large proportion. (4) Generally, the total repair cost, repair costs of structural members and non-structural members increase with the increase of GM intensity. If the retrofitting cost is considered, the economic benefit of the retrofitting is not favorable when a future earthquake has a frequent earthquake intensity, the economic benefit of the retrofitting is still not favorable (but acceptable) when a future earthquake has an occasional earthquake intensity, and the economic benefit of the retrofitting is favorable and can save life when a future earthquake has a rare or larger earthquake intensity.

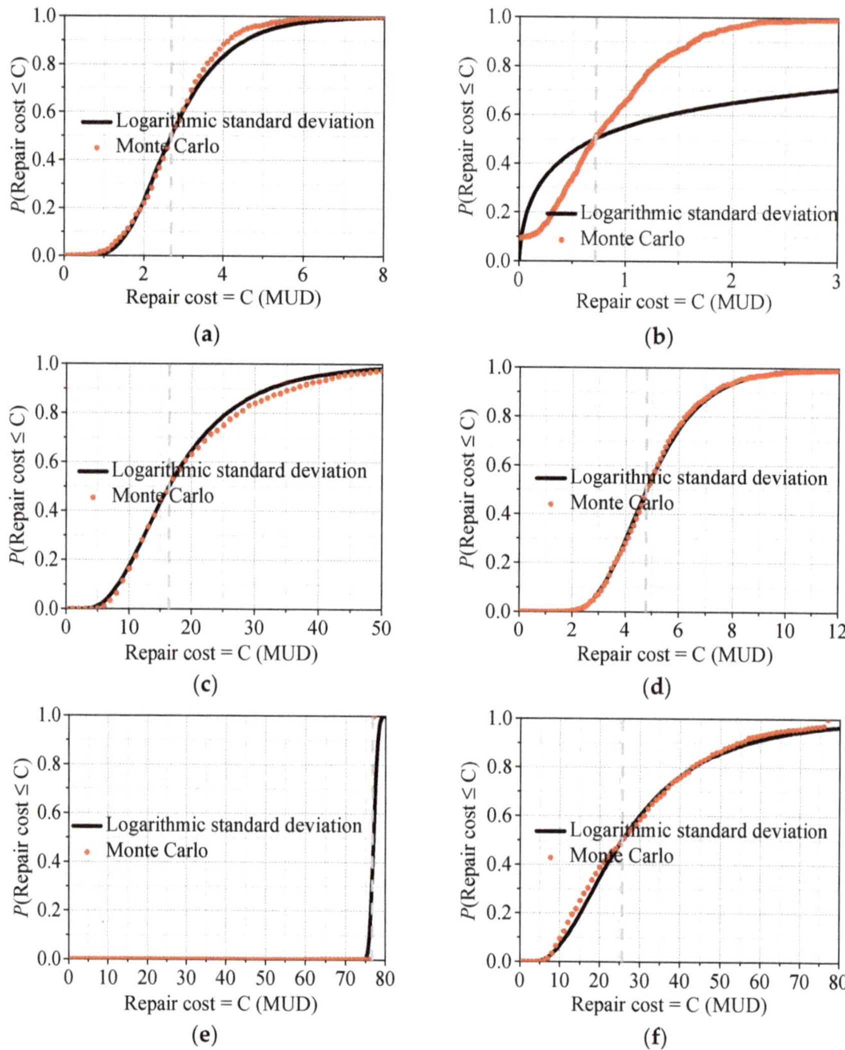

Figure 12. Exceedance probabilities of economic loss under different earthquake intensities: (**a**) 3D original frame, frequent earthquake; (**b**) 3D retrofitted frame, frequent earthquake; (**c**) 3D original frame, occasional earthquake; (**d**) 3D retrofitted frame, occasional earthquake; (**e**) 3D original frame, rare earthquake (frame is collapse in this case); (**f**) 3D retrofitted frame, rare earthquake.

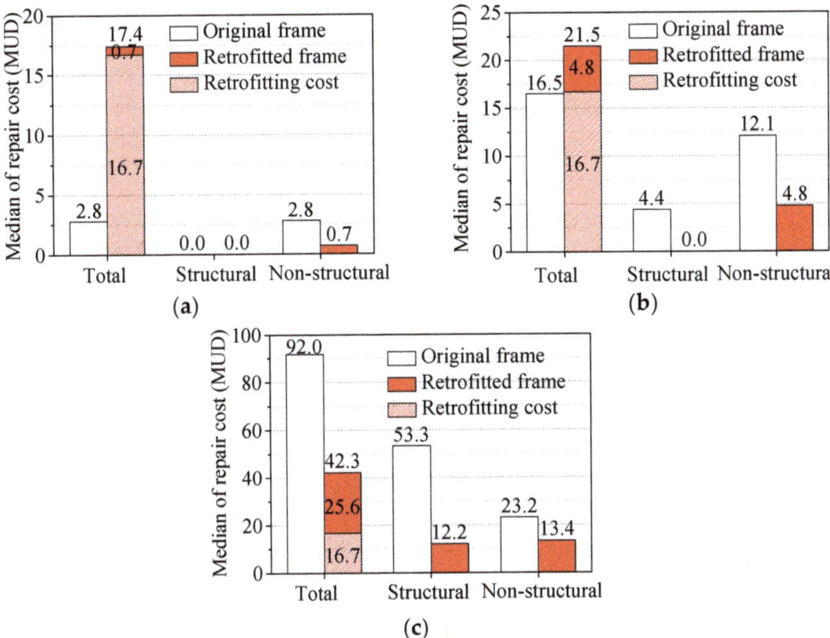

Figure 13. Median values of repair costs under different earthquake intensities: (**a**) Frequent earthquake; (**b**) Occasional earthquake; (**c**) Rare earthquake.

Figure 14a–f compares the repair times between the original frame and retrofitted frame. These figures show that the probabilities are smaller than the specific repair time levels (using the number of workdays as an index). Figure 15 shows the median repair times, which are the values corresponding to a probability of 50% in Figure 14. Table 11 shows the total median repair times of the original frame and the retrofitted frame under a frequent earthquake, occasional earthquake and rare earthquake. The observations are that (1) There are no obvious differences when the two frames are hit by a frequent earthquake, in both cases the repair times are within one week. (2) The repair time for the retrofitted frame is shorter than that of the original frame under an occasional earthquake, requiring 7 workdays for the retrofitted frame and 34 workdays for the original frame. (3) The repair time for the retrofitted frame is obviously shorter than that of the original frame under a rare earthquake, where 53 workdays are required for the retrofitted frame and 402 workdays for the original frame. In this case, the original frame collapses and needs reconstruction. Similar to the calculation of repair cost, demolition and site clearance will increase the structural repair time by up to 20–30%; therefore, the total repair time under rare earthquakes increases by 20%, i.e., 402 × 120% = 482 workdays. Therefore, the ratio between the repair time of retrofitting and the repair time of complete replacement is about 10%. Note that the shorter the repair time to a building, the less the impact of environmental problems.

Table 11. Median repair times of the original frame and retrofitted frame (unit: workday).

Frame	Frequent Earthquake PGA = 0.07 g	Occasional Earthquake PGA = 0.2 g	Rare Earthquake PGA = 0.4 g
3D original frame	5	34	482
3D retrofitted frame	2	7	53

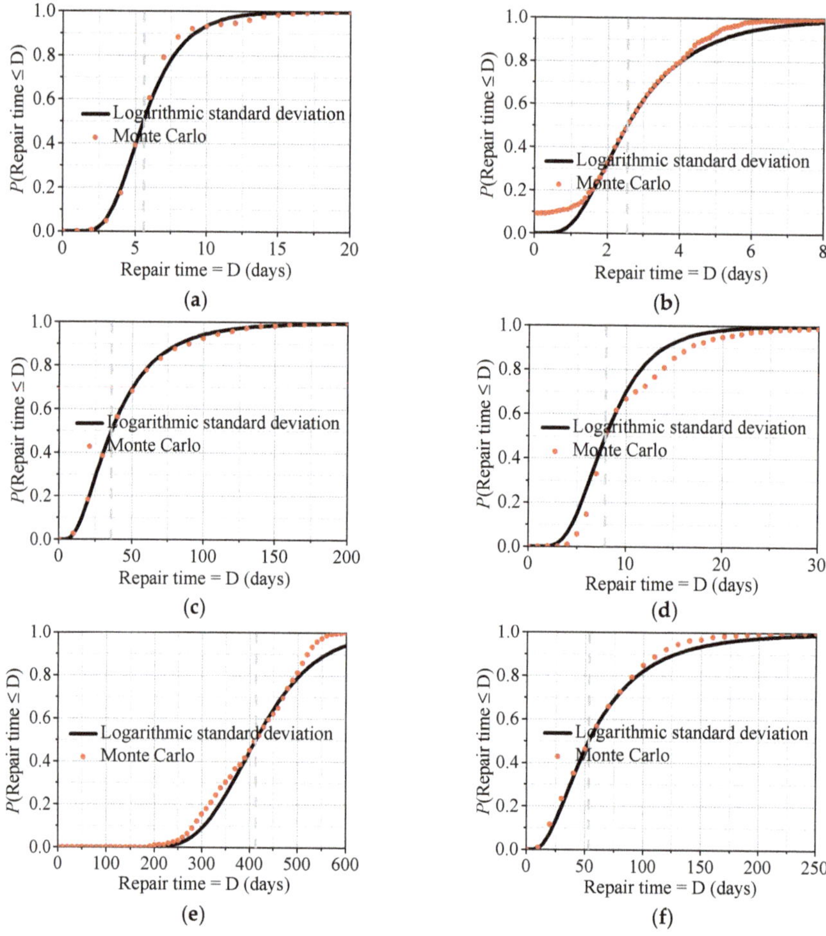

Figure 14. Exceedance probabilities of repair time under different earthquake intensities (Parallel construction): (**a**) 3D original frame, frequent earthquake; (**b**) 3D retrofitted frame, frequent earthquake; (**c**) 3D original frame, occasional earthquake; (**d**) 3D retrofitted frame, occasional earthquake; (**e**) 3D original frame, rare earthquake; (**f**) 3D retrofitted frame, rare earthquake.

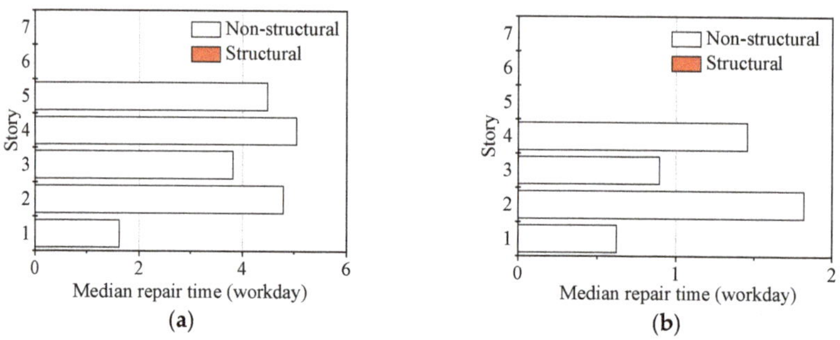

Figure 15. *Cont.*

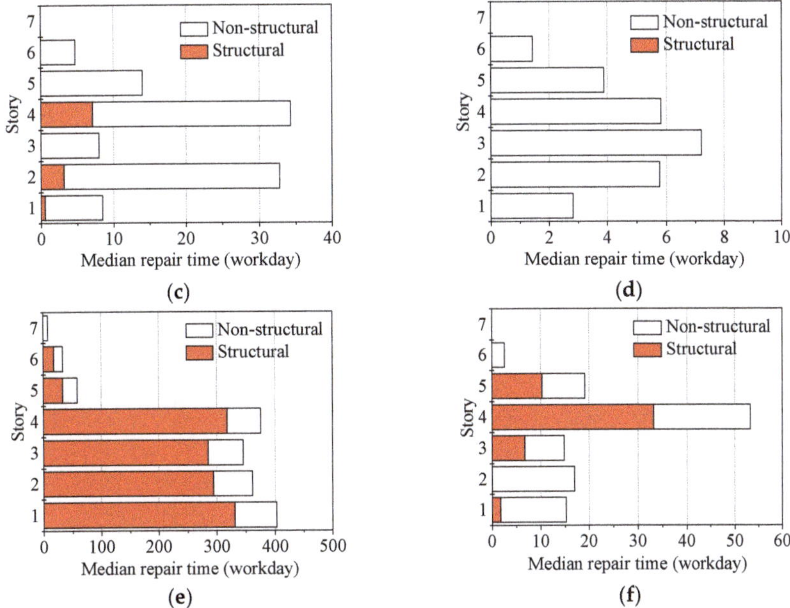

Figure 15. Median values of repair times under different earthquake intensities (parallel construction): (**a**) 3D original frame, frequent earthquake; (**b**) 3D retrofitted frame, frequent earthquake; (**c**) 3D original frame, occasional earthquake; (**d**) 3D retrofitted frame, occasional earthquake; (**e**) 3D original frame, rare earthquake; (**f**) 3D retrofitted frame, rare earthquake.

5. Sustainability Analysis for the Proposed Method

Besides the economic analysis, environmental sustainability is also an important factor nowadays to influencing retrofit planning. Carbon emission is a commonly used index relevant to the ecological environment. The carbon emission calculation for a building can be divided into four stages: material preparation, construction, building use and demolition. The carbon emission model in the life-cycle period is [52]

$$P_{LC} = P_1 + P_2 + P_3 + P_4 \tag{7}$$

where P_{LC} is the carbon emission in the life-cycle period; and P_1, P_2, P_3, and P_4 are the carbon emissions during material preparation, construction, building use and demolition, respectively.

P_1 positively correlates to material usages, their carbon emission factors, and their hauling distances. P_2 positively correlates to the amount of required equipment and their carbon emission factors, and the number of workdays. In this study, the carbon emission during the building use (P_3) for the original frame and for the retrofitted frame is assumed to be almost the same, so P_3 is not further discussed later in the calculation. P_4 positively correlates to $P_1 + P_2$, and so usually $10\%(P_1 + P_2)$ is adopted in the calculation.

The specific estimation process of carbon emission is as follows. P_1 can be calculated using the following equation

$$P_1 = \sum_{i=1}^{n} M_i \times \left(EF_{cl,i} + L_i \times EF_{jt,i} \times 10^{-4} \right) \tag{8}$$

where M_i is the total usage of ith material; $EF_{cl,i}$ is the emission factor of ith material (about 0.2 for steel and 0.15 for concrete [52]); L_i is the hauling distance of the ith material; $EF_{jt,i}$ is the emission factor of ith material during hauling (about 0.07 for railway hauling, 1.65 for highway hauling, and 0.15 for waterway hauling [52]); and n is the number of used material types.

P_2 can be calculated using the following equation

$$P_2 = \sum_{j=1}^{n} N_j \times EF_{sg,j} + R \times 0.012 \tag{9}$$

where N_j is the amount of equipment of type j; $EF_{sg,j}$ is the emission factor of jth equipment; R is the number of workdays; 0.012 is the estimated value of carbon emission of a person per day; and n is the number of equipment types.

The original frame is collapsed and the retrofitted frame is in a repairable state under a rare earthquake. Therefore, the original frame needs complete replacement, which leads to extra demolition (P_4) and much more material usage (P_1) for rebuilding than for retrofitting. According to Equation (8), if the hauling type is the same, P_1 for original frame will be much larger than P_1 for the retrofitted frame, because the M_i for the original frame (in the case of total replacement) will be much larger than that of the retrofitted frame (in the case of repair). It is a difficult thing to estimate how much equipment will be used in the replacement process without a detailed construction organization analysis. However, it is easy to see that more equipment needs to be used in the replacement process than in the repair process. As shown in Table 11, R is 482 days and 53 days for the original frame and retrofitted frame, respectively. Therefore, according to Equation (9), P_2 for original frame will be much larger than P_2 for the retrofitted frame. Since $P_3 = 10\%$ $(P_1 + P_2)$ is adopted in the calculation, P_3 for original frame is larger than P_3 for the retrofitted frame. In summary, if we want to recover the building function after rare earthquakes, the carbon emission for the original frame will be larger than that for the retrofitted frame in the repair process.

6. Conclusions

For a structure with insufficient earthquake resistance, this study proposed a retrofitting method based on an optimum placement of viscous dampers. The following conclusions can be drawn:

(1) The proposed retrofitting method can be applied both to satisfy the requirements of design codes and to enhance the structural collapse-resistant capacity of existing structures. An advantage of the proposed method is that the retrofitted structure has a larger earthquake collapse-resistant capacity compared with an ordinary retrofitting method entailing the same retrofitting cost. For a regular 3D frame, the computational cost is low because a 2D frame model can be adopted to determine the optimum placement pattern of the viscous dampers. Similar to other design methods for the placement of dampers, the proposed method is also a procedure based on a series of time-history analyses. However, for the 6-story RC frame structure used in this study, one time-history analysis commonly only takes about three minutes (for the 2D frame model). Therefore, the computational cost of the proposed method is acceptable.

(2) The economic benefit analysis is conducted to check the convenience of the proposed retrofitting method. The owners can clearly know the economic benefits under different earthquake intensities that may happen in the future. For the 6-story RC frame structure used in this study, during its lifetime cycle, the economic benefit of retrofitting is not favorable if the structure only suffers an earthquake with a frequent earthquake intensity, while the economic benefit of retrofitting is favorable if the structure suffers an earthquake with an occasional earthquake intensity (depending on the building functions which influence the economic loss by downtime/repair time. The example building used in this study is a middle school, meaning that downtime/repair time will not induce much economic loss. However, the economic loss will be higher if it would be a commercial building.), and the economic benefit of retrofitting is significantly favorable if the structure suffers an earthquake with a rare intensity (the ratio between the cost of retrofitting and the cost of complete replacement is about 45%; the ratio between the repair time of retrofitting and the repair time of complete replacement is about 10%); in this case, the retrofitting will save householders' lives in case of earthquake. This economic benefit analysis results can be provided

to householders who may decide whether to perform retrofitting or not. Besides the economic benefit, environmental sustainability is also discussed by way of a carbon emission calculation. Under a rare earthquake, the original frame needs replacement, which leads to more carbon emissions than a retrofitted frame in the repair process.

Author Contributions: Conceptualization and funding acquisition, Writing—review and editing: S.L. Methodology and formal analysis: J.Z.

Funding: This research project was supported by the National Key R&D Program of China (Grant No. 2016YFC0701500), National Natural Science Foundation of China (Grant No. 51578202), and Natural Science Foundation of Heilongjiang (Grant No. YQ2019E021). The financial supports are greatly appreciated by the authors.

Conflicts of Interest: The authors declare no conflict of interest.

References

1. Stone, R. Wenchuan earthquake—Lessons of disasters past could guide Sichuan's revival. *Science* **2008**, *321*, 476. [CrossRef] [PubMed]
2. Yi, F.X.; Tu, Y. An evaluation of the paired assistance to disaster-affected areas program in disaster recovery: The case of the Wenchuan earthquake. *Sustainability* **2018**, *10*, 4483. [CrossRef]
3. *Code for Seismic Design of Buildings, GB50010-2010*; Ministry of Housing and Urban-Rural Development of P. R.: Beijing, China, 2010.
4. Smyth, A.W.; Altay, G.; Deodatis, G.; Erdik, M.; Franco, G.; Gulkan, P.; Kunreuther, H.; Lus, H.; Mete, E.; Seeber, N.; et al. Probabilistic benefit-cost analysis for earthquake damage mitigation: Evaluating measures for apartment houses in Turkey. *Earthq. Spectra* **2004**, *20*, 171–203. [CrossRef]
5. Di Sarno, L.; Manfredi, G. Seismic retrofitting with buckling restrained braces: Application to an existing non-ductile RC framed building. *Soil Dyn. Earthq. Eng.* **2010**, *30*, 1279–1297. [CrossRef]
6. Masi, A.; Santarsiero, G.; Chiauzzi, L. Development of a seismic risk mitigation methodology for public buildings applied to the hospitals of Basilicata region (Southern Italy). *Soil Dyn. Earthq. Eng.* **2014**, *65*, 30–42. [CrossRef]
7. Xiang, P.; Nishitani, A. Seismic vibration control of building structures with multiple tuned mass damper floors integrated. *Earthq. Eng. Struct. Dyn.* **2014**, *43*, 909–925. [CrossRef]
8. Johnson, J.G.; Pantelides, C.P.; Reaveley, L.D. Nonlinear rooftop tuned mass damper frame for the seismic retrofit of buildings. *Earthq. Eng. Struct. Dyn.* **2015**, *44*, 299–316. [CrossRef]
9. Ghiani, C.; Linul, E.; Porcu, M.C.; Marsavina, L.; Movahedi, N.; Aymerich, F. Metal foam-filled tubes as plastic dissipaters in earthquake-resistant steel buildings. *IOP Conf. Ser. Mater. Sci Eng.* **2018**, *416*, 012051. [CrossRef]
10. Porcu, M.C.; Vielma, J.C.; Panu, F.; Aguilar, C.; Curreli, G. FRP seismic retrofit of existing buildings led by dynamic non-linear analyses. *Int. J. Saf. Secur. Eng.* **2019**, *9*, 201–212. [CrossRef]
11. Porcu, M.C. Partial floor mass isolation to control seismic stress in framed buildings. *Int. J. Saf. Secur. Eng.* **2019**, *9*, 157–165. [CrossRef]
12. Pekcan, G.; Mander, J.B.; Chen, S.S. *Design and Retrofit Methodology for Building Structures with Supplemental Energy Dissipating Systems*; 99-0021; Technical Report MCEER: Buffalo, NY, USA, 1999.
13. Uriz, P.; Whittaker, A.S. Retrofit of pre-Northridge steel moment-resisting frames using fluid viscous dampers. *Struct. Des. Tall Spec. Buil.* **2001**, *10*, 371–390. [CrossRef]
14. Silvestri, S.; Trombetti, T. Physical and numerical approaches for the optimal insertion of seismic viscous dampers in shear-type structures. *J. Earthq. Eng.* **2007**, *11*, 787–828. [CrossRef]
15. Impollonia, N.; Palmeri, A. Seismic performance of buildings retrofitted with nonlinear viscous dampers and adjacent reaction towers. *Earthq. Eng. Struct. Dyn.* **2018**, *47*, 1329–1351. [CrossRef]
16. Karavasilis, T.L. Assessment of capacity design of columns in steel moment resisting frames with viscous dampers. *Soil Dyn. Earthq. Eng.* **2016**, *88*, 215–222. [CrossRef]
17. Kim, J.; Choi, H.; Min, K.W. Performance-based design of added viscous dampers using capacity spectrum method. *J. Earthq. Eng.* **2003**, *7*, 1–24. [CrossRef]
18. Lin, Y.Y.; Tsai, M.H.; Hwang, J.S.; Chang, K.C. Direct displacement-based design for building with passive energy dissipation systems. *Eng. Struct.* **2003**, *25*, 25–37. [CrossRef]

19. Habibi, A.; Chan, R.W.K.; Albermani, F. Energy-based design method for seismic retrofitting with passive energy dissipation systems. *Eng. Struct.* **2013**, *46*, 77–86. [CrossRef]

20. Palermo, M.; Silvestri, S.; Landi, L.; Gasparini, G.; Trombetti, T. "A direct five-step procedure" for the preliminary seismic design of buildings with added viscous dampers. *Eng. Struct.* **2018**, *173*, 933–950. [CrossRef]

21. Zhou, Y.; Lu, X.L.; Weng, D.G.; Zhang, R.F. A practical design method for reinforced concrete structures with viscous dampers. *Eng. Struct.* **2012**, *39*, 187–198. [CrossRef]

22. Vesna, T.; Stephen, M.; Mary, C.C. Comparative life-cycle cost and performance analysis of structural systems for buildings. In Proceedings of the NCEE 2014—10th U.S. National Conference on Earthquake Engineering: Frontiers of Earthquake Engineering, Anchorage, AK, USA, 21–25 July 2014.

23. Zhang, R.H.; Soong, T.T. Seismic Design of Viscoelastic Dampers for Structural Applications. *J. Struct. Eng.-ASCE* **1992**, *118*, 1375–1392. [CrossRef]

24. Singh, M.P.; Moreschi, L.M. Optimal seismic response control with dampers. *Earthq. Eng. Struct. Dyn.* **2001**, *30*, 553–572. [CrossRef]

25. Cimellaro, G.P. Simultaneous stiffness-damping optimization of structures with respect to acceleration, displacement and base shear. *Eng. Struct.* **2007**, *29*, 2853–2870. [CrossRef]

26. Apostolakis, G.; Dargush, G.F. Optimal seismic design of moment-resisting steel frames with hysteretic passive devices. *Earthq. Eng. Struct. Dyn.* **2010**, *39*, 355–376. [CrossRef]

27. Aydin, E. Optimal damper placement based on base moment in steel building frames. *J. Constr. Steel Res.* **2012**, *79*, 216–225. [CrossRef]

28. Martinez, C.A.; Curadelli, O.; Compagnoni, M.E. Optimal design of passive viscous damping systems for buildings under seismic excitation. *J. Constr. Steel Res.* **2013**, *90*, 253–264. [CrossRef]

29. Shin, H.; Singh, M.P. Minimum failure cost-based energy dissipation system designs for buildings in three seismic regions—Part I: Elements of failure cost analysis. *Eng. Struct.* **2014**, *74*, 266–274. [CrossRef]

30. Pollini, N.; Lavan, O.; Amir, O. Minimum-cost optimization of nonlinear fluid viscous dampers and their supporting members for seismic retrofitting. *Earthq. Eng. Struct. Dyn.* **2017**, *46*, 1941–1961. [CrossRef]

31. Parcianello, E.; Chisari, C.; Amadio, C. Optimal design of nonlinear viscous dampers for frame structures. *Soil Dyn. Earthq. Eng.* **2017**, *100*, 257–260. [CrossRef]

32. Lavan, O.; Amir, O. Simultaneous topology and sizing optimization of viscous dampers in seismic retrofitting of 3D irregular frame structures. *Earthq. Eng. Struct. Dyn.* **2014**, *43*, 1325–1342. [CrossRef]

33. Wang, S.; Mahin, S.A. High-performance computer-aided optimization of viscous dampers for improving the seismic performance of a tall building. *Soil Dyn. Earthq. Eng.* **2018**, *113*, 454–461. [CrossRef]

34. De Domenico, D.; Ricciardi, G. Earthquake protection of structures with nonlinear viscous dampers optimized through an energy-based stochastic approach. *Eng. Struct.* **2019**, *179*, 523–539. [CrossRef]

35. Li, S.; Yu, B.; Gao, M.M.; Zhai, C.H. Optimum seismic design of multi-story buildings for increasing collapse resistant capacity. *Soil Dyn. Earthq. Eng.* **2019**, *116*, 495–510. [CrossRef]

36. Li, S.; Kose, M.M.; Shan, S.D.; Sezen, H. Modeling methods for collapse analysis of reinforced concrete frames with infill walls. *J. Struct. Eng.* **2019**, *145*, 04019011. [CrossRef]

37. Open System for Earthquake Engineering Simulation (OpenSees). 2016. Available online: http://opensees.berkeley.edu/wiki/index.php/Command_Manual (accessed on 9 January 2018).

38. Li, S.; Zuo, Z.; Zhai, C.; Xu, S.; Xie, L. Shaking table test on the collapse process of a three-story reinforced concrete frame structure. *Eng. Struct.* **2016**, *118*, 156–166. [CrossRef]

39. *Code for Seismic Design of Building, GB50011-2010*; Ministry of Housing and Urban Rural Development: Beijing, China, 2010. (In Chinese)

40. *Seismic Evaluation and Retrofit of Existing Buildings, ASCE/SEI 7–16*; American Society of Civil Engineers: Reston, VA, USA, 2016.

41. Part 1: General rules, seismic action and rules for buildings. In *Eurocode 8. Design of Structures for Earthquake Resistance*; European Committee for Standardization: Brussels, Belgium, 2004.

42. Vamvatsikos, D.; Cornell, C.A. Incremental dynamic analysis. *Earthq. Eng. Struct. Dyn.* **2002**, *31*, 491–514. [CrossRef]

43. *FEMA 350, Recommended Seismic Design Criteria for New Steel Moment-Frame Buildings*; Applied Technology Council: Redwood City, CA, USA, 2000.

44. *FEMA 356, Prestandard and Commentary for the Seismic Rehabilitation of Buildings*; Applied Technology Council: Redwood City, CA, USA, 2000.

45. Benavent-Climent, A.; Cahis, X.; Vico, J.M. Interior wide beam-column connections in existing RC frames subjected to lateral earthquake loading. *Bull. Earthq. Eng.* **2010**, *8*, 401–420. [CrossRef]

46. Benavent-Climent, A.; Cahis, X.; Zahran, R. Exterior wide beam-column connections in existing RC frames subjected to lateral earthquake loads. *Eng. Struct.* **2009**, *31*, 1414–1424. [CrossRef]

47. Masi, A.; Santarsiero, G. Seismic tests on RC building exterior joints with wide beams. *Adv. Mater. Res.* **2013**, *787*, 771–777. [CrossRef]

48. Zareian, F.; Krawinkler, H. Assessment of probability of collapse and design for collapse safety. *Earthq. Eng. Struct. Dyn.* **2007**, *36*, 1901–1914. [CrossRef]

49. *FEMA P-58, Seismic Performance Assessment of Buildings, Volume 1—Methodology*; Applied Technology Council: Redwood City, CA, USA, 2012.

50. China Engineering Cost Network. 2019. Available online: http://www.cecn.gov.cn/ (accessed on 2 January 2019).

51. *TY 01-89-2016. Quota for Days of Construction of Structures*; Ministry of Housing and Urban Rural Development: Beijing, China, 2016. (In Chinese)

52. Dong, K.H. Study on Carbon Emission Caculation from the Life-Cycle of Large-Scale Complex Building in the South. Master's Thesis, South China University of Technology, Guangzhou, China, 2018.

Article

Finite Element Steady-State Vibration Analysis Considering Frequency-Dependent Soil-Pile Interaction

Seung-Han Song [1] and Sean Seungwon Lee [2,*]

[1] Principal Engineer, CSA Department, Worley, Houston, TX 77072, USA; seung.song@worley.com
[2] Department of Earth Resources and Environmental Engineering, Hanyang University, Seoul 04763, Korea
* Correspondence: seanlee@hanyang.ac.kr; Tel.: +82-02-2220-2243

Received: 22 November 2019; Accepted: 6 December 2019; Published: 9 December 2019

Abstract: The vibration response of equipment foundation structures is not only affected by the structural stiffness and mass, but also greatly influenced by the degree of a soil-foundation structural interaction. Furthermore, the vibratory performance of equipment foundation structures supported by pile systems largely depends on the soil-pile dynamic stiffness and damping, which are variable in nature within the speed range that machines operate at. This paper reviews a method for evaluating effective soil-pile stiffness and damping that can be computed by Novak's method or by commercial software (DYNA6, University of Western Ontario). A series of Finite Element (FE) time history and steady-state analyses using SAP2000 have been performed to examine the effects of dynamic soil-pile-foundation interaction on the vibration performance of equipment foundations, such as large compressor foundations and steam/gas turbine foundations. Frequency-dependent stiffness is estimated to be higher than frequency-independent stiffness, in general, and, thus, affects the vibration calculation. This paper provides a full-spectrum steady-state vibration solution, which increases the reliability of the foundation's structural design.

Keywords: FEM (Finite Element Method); DYNA6; soil-structure interaction; soil-pile dynamic stiffness

---

1. Introduction

The importance of foundational dynamic stiffness and damping in the vibration's assessment has been addressed in many literatures such as Novak's method [1], Kausel's approach [2], and Roesset's research [3].

The conventional method for evaluating soil stiffness and damping is based on the classical theory of vibrations of a disk supported on top of an elastic half-space [4]. This theoretical solution is limited to few simple foundation configurations and soil profiles. The applications of the theory often overestimate the degree of soil damping, and, thus, underestimates the vibration amplitude. Field experiences show that the beneficial effects of radiation damping in mitigating foundational vibrations, as predicted by the elastic half-space vibration theory, may not always be effective [5]. Such cases include large compressor foundations, combustion turbine generator/steam turbine generator (CTG/STG) table-top foundations, and other large foundational structures that support high-speed (30 Hz or higher) vibratory equipment. In special soil media, such as rock stratum found very close to the bottom of the foundation, the radiation damping from the theory may be impractical to vibration analyses. This minimal radiation damping is due to the elastic rebounding of waves at the soft soil-rock interface, which is different from the theoretical solution. Currently, a few modelling guidelines [5] exist for assessing such a soil-structure interaction damping on foundation vibrations using the Finite Element Method. This paper proposes several state-of-the-art methods for evaluating effective soil damping that could be modelled in FE (Finite Element) software i.e., SAP2000 and GTSTRUDL. For the formulation of modal damping ratios in the dynamic equation of motion, two different approaches, i.e., direct use of effective damping values and proportional damping (either mass-proportional or stiffness-proportional damping, or both) are employed in this study.

The frequency-dependent interaction effect [6,7] is accomplished by numerical simulations utilizing DYNA6 [8] and is incorporated in the frequency domain steady-state vibration analysis in this paper.

2. Dynamic Soil Profile

The idealized soil profile located beneath the sample CTG and STG foundations is depicted in Figure 1. The dynamic characteristics of soil properties, i.e., shear wave velocity, shear modulus, unit weight, and Poisson ratio for each layer, are summarized. The frequency-independent soil stiffness and damping are formulated by the Elastic Half-Space Method [4] and is used initially in the analysis. The consideration of soil uncertainty is not accounted for in the sample analyses to minimize the computational effort.

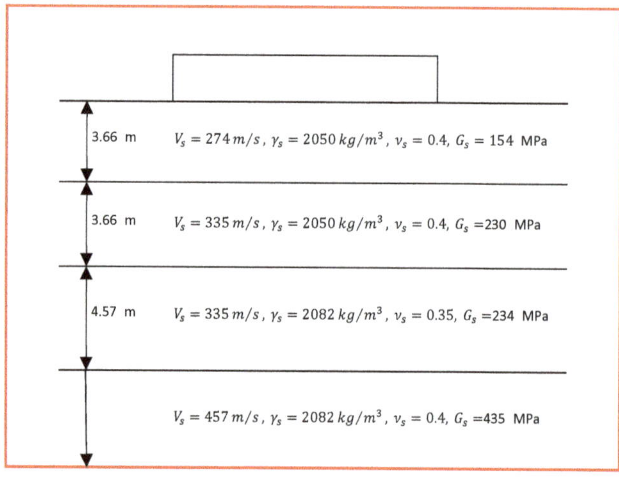

Figure 1. Dynamic soil profiles under CTG/STG Foundations V_s: Shear wave velocity of soil, γ_s: Unit weight of soil, v_s: Poisson ratio of soil, G_s: Shear modulus of soil.

3. Analytical CTG/STG FE Models

The FE models of the sample CTG and STG foundations were developed with two commercial programs, SAP2000 [9] and GTSTRUDL [10]. The two programs provide eight-node solid element equipped with different shape functions, i.e., bending improved and non-bending improved. The CTG foundation (5.49 m wide, 18.29 m long, and 1.52 m thick) supports a set of turbine and generator equipment, which weighs a total of 210.01 t. Both the SAP2000 and GTSTRUDL CTG models consist of 624 eight-node solid elements and 972 joints (2916 degrees of freedom). The STG foundation (18.90 m wide, 36.58 m long, and 13.11 m tall) also supports a set of turbine and generator equipment together with a heavy condenser. The weight of the turbine and generator is estimated to be 857.29 t and the condenser is assumed to weigh 578.33 t.

The STG foundation models consist of 3828 eight-node solid elements and 6335 joints (19,005 degrees of freedom). The SAP2000 models additionally include nine incompatible bending shape functions in the eight-node solid elements so that the bending behaviour of the element can be significantly improved, which is not featured in GTSTRUDL models. The eight-node solid elements used in GTSTRUDL models adopt the 'IPSL (Isoparametric Solid Linear Displacement)' elements.

The dynamic FE models are illustrated in Figures 2 and 3. The material property of both CTG and STG foundations is assumed to be 27.6 MPa cylindrical concrete strength with 24,856 MPa Young's modulus. The unit weight of concrete is of typical 2403 kg/m^3.

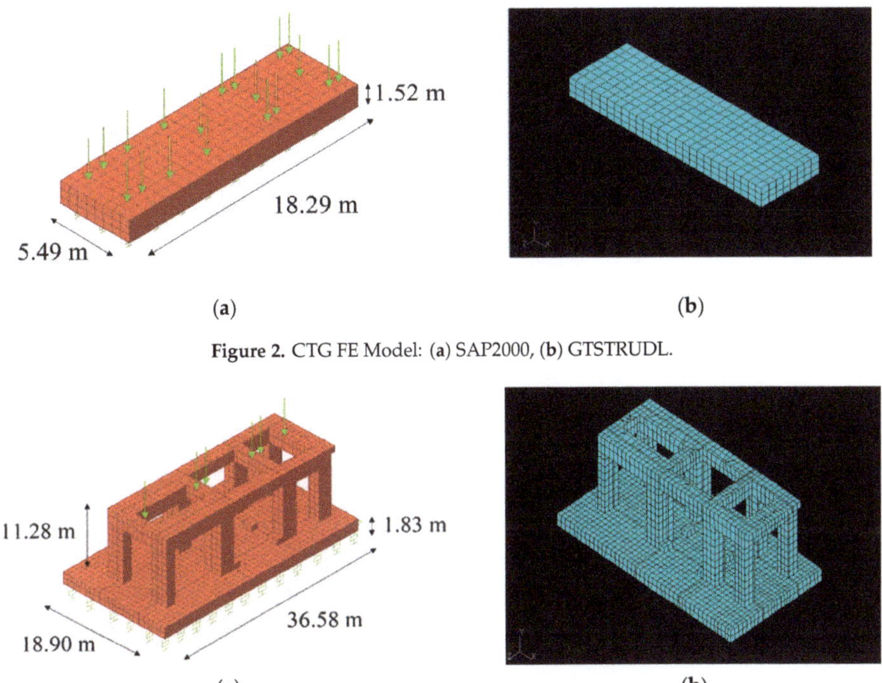

Figure 2. CTG FE Model: (**a**) SAP2000, (**b**) GTSTRUDL.

Figure 3. STG FE Model: (**a**) SAP2000, (**b**) GTSTRUDL.

4. Dynamic Unbalanced Forces

The dynamic unbalanced forces at 60 Hz excitation frequency are applied to the CTG and STG FE models. The CTG foundation force amplitudes and locations are all designated by the CTG vendor. Two scenario-based phase angle conditions are considered in the analysis. One is VTX60A, which is the case when all X, Y, and Z directional loads on the turbine and generator are in-phase. The other is VTX60B, which is the case when all the loads on the turbine are 180° out-of-phase with the loads on the generator.

The STG foundation force amplitudes and phase angle cases are plotted in Figure 4. A total of eight load cases were analysed, to simulate meaningful phase angle cases, i.e., V60A, H60A, V60B, H60B, V60C, H60C, V60D, and H60D. V60A and H60A are the vertical and transverse load cases, respectively, where all the loads are in-phase. V60B and H60B are also vertical and transverse load cases where unbalanced forces at bearings 5 and 6 are 180° out-of-phase with those at bearings 1, 2, 3, and 4. V60C and H60C are also vertical and transverse load cases where unbalance forces at bearings 1 and 2 are 180° out-of-phase with those at bearings 3, 4, 5, and 6. V60D and H60D are also vertical and transverse load cases where unbalanced forces at bearings 3 and 4 are 180° out-of-phase with those at bearings 1, 2, 5, and 6. The unbalanced force amplitudes at 60 Hz are assumed to be 1.81 t at bearings 1 and 2, 5.67 t at bearings 3 and 4, and 4.54 t at bearings 5 and 6, as per the vendor recommendation. The location of bearing, i.e., point of unbalanced force, is specified by the equipment vendor. However, in general, it is two ends of each combustion/steam turbine generator segment at the foundation center in the axial direction.

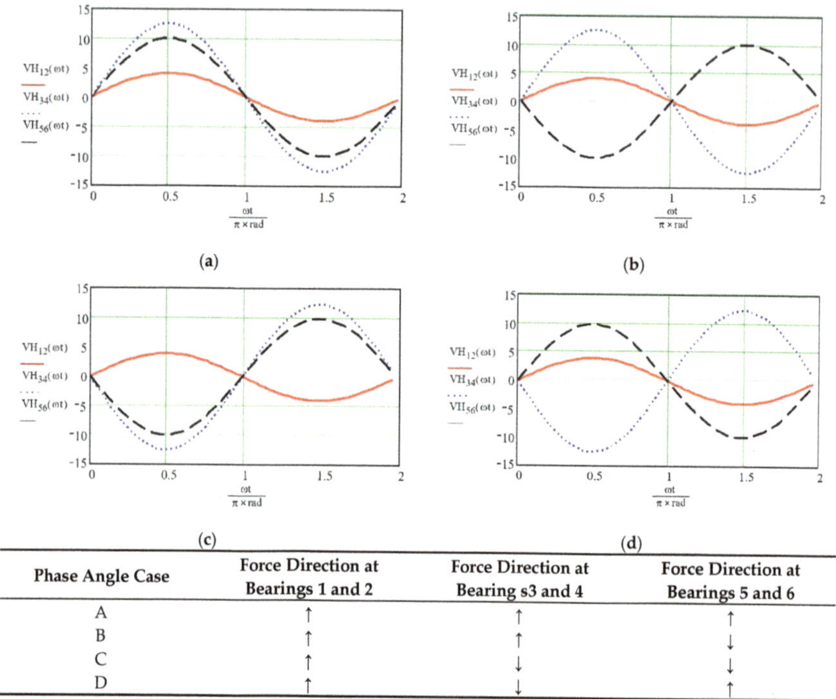

Phase Angle Case	Force Direction at Bearings 1 and 2	Force Direction at Bearing s3 and 4	Force Direction at Bearings 5 and 6
A	↑	↑	↑
B	↑	↑	↓
C	↑	↓	↓
D	↑	↓	↑

Figure 4. STG Foundation Dynamic Unbalance Force at 60 Hz: (**a**) V60A, H60A, (**b**) V60B, H60B, (**c**) V60C, H60C, (**d**) V60D, H60D.

5. Analysis Methods for CTG and STG Foundational Vibration Assessment

A number of analysis methods can be proposed for the vibration assessment of the CTG and STG foundations. The critical issue is how to model soil radiation damping, and, therefore, how much the contribution of soil damping would affect the foundation's vibration response at 60 Hz. Two dynamic analysis methods are considered: modal superposition time history and direct integration time history. In the modal superposition time history analysis, five different soil damping cases are considered: (1) a direct soil damping coefficient from the elastic half-space solution plus 2% concrete material damping, (2) a direct soil damping coefficient from the elastic half-space solution with a 20% cut-off from the EPRI (Electric Power Research Institute) plus 2% concrete material damping, (3) all modal damping at 2%, (4) all modal damping at 4%, and (5) all modal damping at 10%. The modal damping of the (1) and (2) models is, therefore, computed by SAP2000 systematically depending on the modal deformation shapes, i.e., the energy is proportional at each mode and at the interface of the foundation and the soil.

For the direct integration time history analysis, the conventional Newmark β method (β = 0.25 assumed) [11] is used throughout the analyses and three sets of proportional damping parameters (α: mass proportional damping and β: stiffness-proportional damping) [12] are considered: (1) ξ = 4% at 1st modal frequency and ξ = 4% at 60 Hz, (2) ξ = 10% at 1st modal frequency and ξ = 4% at 60 Hz, and (3) ξ = 20% at the first modal frequency and ξ = 4% at 60 Hz. The general form of α and β based on two frequency points is given by the following equations.

$$\alpha := 2 \cdot \frac{\xi_1 \cdot \omega_2 - \xi_2 \cdot \omega_1}{\left(\frac{\omega_2}{\omega_1} - \frac{\omega_1}{\omega_2}\right)} \cdot sec \tag{1}$$

$$\beta := 2 \cdot \frac{\xi_1 \cdot \omega_1 - \xi_2 \cdot \omega_2}{\omega_1{}^2 - \omega_2{}^2} \cdot \frac{1}{sec} \qquad (2)$$

in which ξ_1 and ξ_2 are damping ratios that correspond to the two frequencies, ω_1 and ω_2, respectively.

It should be noted that both dynamic analysis methods are approximate in terms of their solution techniques. For instance, the modal superposition solution (steady-state part only) depends on only the diagonal terms of the damping matrix but does not account for non-diagonal damping terms in the dynamic equation of motion. On the other hand, the direct integration method utilizes every term in the damping matrix. From this point of view, for the particular case where coupled modes contribute significantly to the total response, the modal superposition solution may not result in an accurate response representation. On the contrary, the direct integration method depends on the time increment, and the two frequency point proportional dampings may lead to an approximate response as well. Therefore, foundation engineers must pay close attention to the selection of damping parameters and analysis methods. A summary table of the applied analysis methods for the vibration assessment of the CTG and STG foundations is given in Table 1.

Table 1. Analysis methods for the CTG and STG foundation vibration assessment.

Analysis Method (Damping)		CTG Foundation	STG Foundation
Modal Superposition Time History Analysis	Direct Soil Damping Coefficient from Elastic Half Space Solution Plus 2% Concrete Material Damping	SAP2000 (20 Modes)	SAP2000 (120 Modes)
	Direct Soil Damping Coefficient from Elastic Half Space Solution with 20% Cutoff by EpRI Plus 2% Concrete Material Damping		
	Modal Damping for All Modes — 2%	SAP2000, GTSTRUDL (20 Modes)	SAP2000, GTSTRUDL (120 Modes)
	Modal Damping for All Modes — 4%		
	Modal Damping for All Modes — 10%		
Direct Integration Time History Analysis (Newmark β=0.25)	Rayleigh Damping — ξ=4% at ω_{1st}, ξ=4% at 60Hz	GTSTRUDL	N/A *
	Rayleigh Damping — ξ=10% at ω_{1st}, ξ=4% at 60Hz		
	Rayleigh Damping — ξ=20% at ω_{1st}, ξ=4% at 60Hz		

* It requires significant computation memory and time.

6. Steady State Time History and Direct Integration Time History Response

The steady-state time history solutions by modal superposition analysis were executed by both SAP2000 and GTSTRUDL. To compute the steady-state modal solution, eigenvalue/eigenvector analysis must be performed before conducting forced vibrational analysis. A total of 20 and 120 modes were considered in the CTG and STG foundation models, respectively, and the higher modes, i.e., more than 120% of excitation frequency, should be included in the modal solution computation. Then, modal combination using either the SRSS (square root of the sum of squares) or CQC (complete quadratic combination) method is performed to compute the total solution. To simplify, only one cycle response was considered using a periodic time history motion.

The direct integration analysis of CTG foundations was also performed with a full damping matrix. Twenty integration points per cycle for a total of 50 cycles were used for the convergence of the steady-state solution. Results of the time history response are illustrated in Figures 5–7.

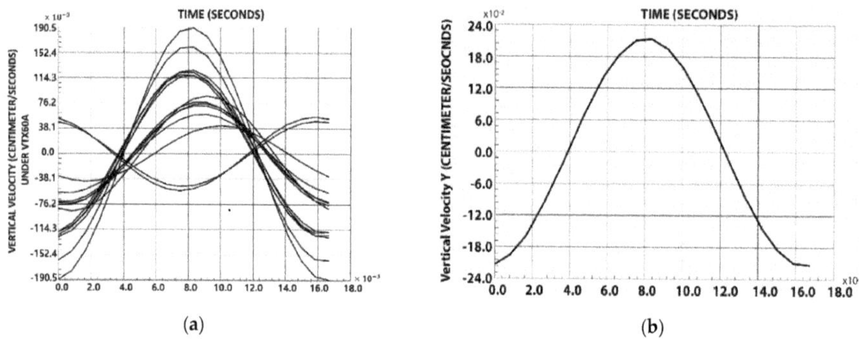

Figure 5. Steady-state time history velocity response (CTG Foundation under VTX60A Load): (**a**) SAP2000, (**b**) GTSTRUDL.

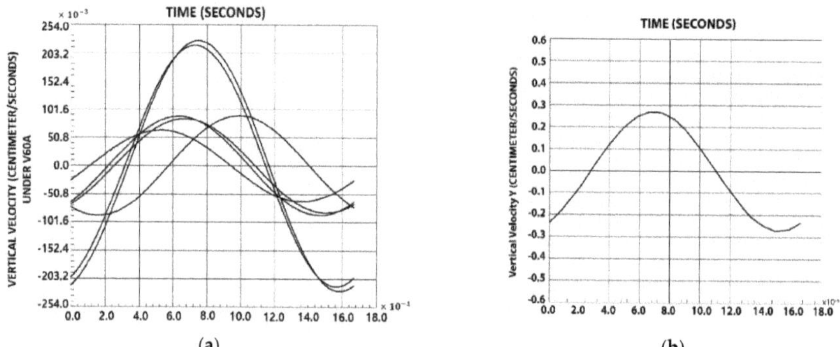

Figure 6. Steady-State Time History Velocity Response (STG Foundation under V60A Load): (**a**) SAP2000 and (**b**) GTSTRUDL.

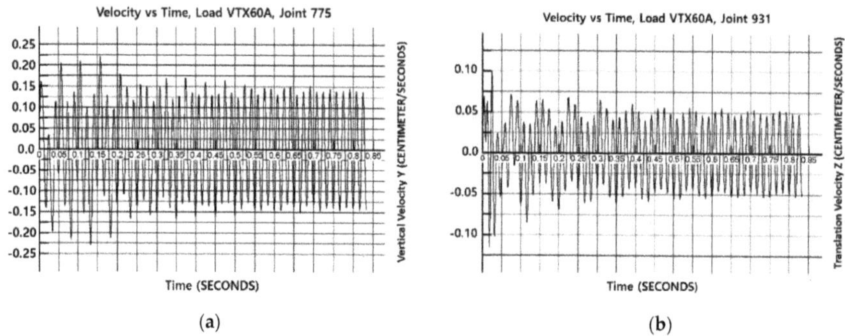

Figure 7. Direct Integration Time History Velocity Response. (CTG Foundation under VTX60A Load, 50 Cycles): (**a**) Max.Velocity a Joint 775, (**b**) Max.Velocity at Joint 931.

A complete maximum steady-state (or permanent) velocity response at the bearings of CTG and STG foundations are tabulated in the following four tables (Tables 2–5). A separate table is provided for each directional velocity, i.e., vertical and transverse.

Table 2. Maximum vertical velocity steady-state eesponse at bearings of the CTG foundation (cm/s).

Damping Method		VTX60A	VTX60B
Direct soil damping value with 2% material damping SAP 2000		0.132	0.155
Direct soil damping value with 20% cutoff by EPRI 2% material damping SAP 2000		0.178	0.302
Modal damping 2%	SAP 2000	0.191	0.351
	GTSTRUDL	0.213	0.442
Modal damping 4%	SAP 2000	0.185	0.333
	GTSTRUDL	0.201	0.391
Modal damping 10%	SAP 2000	0.170	0.274
	GTSTRUDL	0.170	0.290
Rayleigh damping (α = 5.3174, β = 0.00017) GTSTRUDL		0.152	0.127
Rayleigh damping (α = 15.4648, β = 0.0001) GTSTRUDL		0.147	0.122
Rayleigh damping (α = 32.3771, β = 0.0) GTSTRUDL		0.142	0.117

Table 3. Maximum transverse velocity steady-state response at bearings of CTG foundation (cm/s).

Damping Method		VTX60A	VTX60B
Direct soil damping value with 2% material damping SAP 2000		0.127	0.112
Direct soil damping value with 20% cutoff by EPRI 2% material damping SAP 2000		0.132	0.145
Modal damping 2 %	SAP 2000	0.135	0.163
	GTSTRUDL	0.127	0.165
Modal damping 4%	SAP 2000	0.132	0.152
	GTSTRUDL	0.124	0.150
Modal damping 10%	SAP 2000	0.124	0.122
	GTSTRUDL	0.117	0.114
Rayleigh damping (α = 5.3174, β = 0.00017) GTSTRUDL		0.056	0.033
Rayleigh damping (α = 15.4648, β = 0.0001) GTSTRUDL		0.053	0.030
Rayleigh damping (α = 32.3771, β = 0.0) GTSTRUDL		0.051	0.030

Table 4. Maximum vertical velocity steady-state response at bearings of STG foundation (cm/s).

Damping Method		V60A	H60A	V60B	H60B	V60C	H60C	V60D	H60D
Direct soil damping value with 2% material damping SAP 2000		0.198	0.008	0.163	0.008	0.208	0.008	0.178	0.008
Direct soil damping value with 20% cutoff by EPRI plus 2% material damping SAP 2000		0.224	0.053	0.185	0.028	0.229	0.043	0.188	0.020
Modal damping 2%	SAP 2000	0.226	0.124	0.175	0.061	0.221	0.086	0.160	0.061
	GTSTRUDL	0.272	0.147	0.241	0.069	0.310	0.127	0.262	0.081
Modal damping 4%	SAP 2000	0.226	0.058	0.191	0.028	0.234	0.051	0.196	0.025
	GTSTRUDL	0.267	0.074	0.213	0.033	0.300	0.066	0.236	0.033
Modal damping 10%	SAP 2000	0.226	0.013	0.175	0.008	0.239	0.013	0.196	0.008
	GTSTRUDL	0.244	0.015	0.173	0.008	0.262	0.015	0.203	0.010

Table 5. Maximum transverse velocity steady-state response at bearings of STG foundation (cm/s).

Damping Method		V60A	H60A	V60B	H60B	V60C	H60C	V60D	H60D
Direct soil damping value with 2% material damping SAP 2000		0.008	0.094	0.010	0.069	0.005	0.089	0.008	0.074
Direct soil damping value with 20% cutoff by EPRI 2% material damping SAP 2000		0.028	0.140	0.030	0.079	0.038	0.135	0.018	0.114
Modal damping 2%	SAP 2000	0.061	0.193	0.071	0.104	0.089	0.168	0.046	0.221
	GTSTRUDL	0.056	0.168	0.053	0.056	0.069	0.168	0.064	0.178
Modal damping 4%	SAP 2000	0.030	0.150	0.028	0.081	0.038	0.137	0.018	0.135
	GTSTRUDL	0.030	0.145	0.025	0.061	0.030	0.132	0.023	0.132
Modal damping 10%	SAP 2000	0.010	0.114	0.008	0.074	0.010	0.104	0.008	0.084
	GTSTRUDL	0.010	0.112	0.008	0.071	0.010	0.107	0.005	0.076

7. Soil-Pile Interaction Study and Frequency-Dependent Stiffness and Damping

Another sample model has been used to investigate the effect of soil-pile-foundation interaction on the vibration performance of the equipment foundation. The material property of the foundation is assumed to be 27.6 MPa cylindrical concrete strength with 24.856 MPa Young's modulus. The unit weight of concrete is of typical 2403 kg/m^3.

A total of 2706 joints, 70 links, and 1170 solid elements were used in the example model (Figure 8, 3D solid element model with link element). The 3-D solid elements are expected to predict displacement/velocity response with higher accuracy, and the behaviour of the pile foundation with higher efficiency. This model used only three translational degrees of freedom at each joint. A direct soil-pile interaction study has been conducted using the DYNA6 program. A total of 70 CIP (600 mm diameter, 18 m long Cast-In-Place with 34.5 MPa cylindrical concrete strength) piles are installed beneath the large compressor foundation and are modelled by the FE link elements. Two layers of soil media, with shear wave velocities equal to Vs = 300 m/s for the upper 3 m of soil and V$_b$ = 538 m/s below the soil, are expected to interact with the CIP piles, including a weak zone (Figure 9). The weak zone parameters are assumed to have a Poisson ratio of 0.3 and material damping of 5%. The compressor vendor has set the single machine speed at 66.5 Hz. Thus, all the foundational vibration responses up to 120% (80 Hz) of the operating speed were calculated and considered for resonance checking at any frequency. For a reference purpose, an eigenvalue analysis using frequency independent stiffness was performed. The fundamental modal period returned was 0.395 s (horizontal mode coupled with rocking) and 0.099 s (vertical mode), as shown in Figure 10.

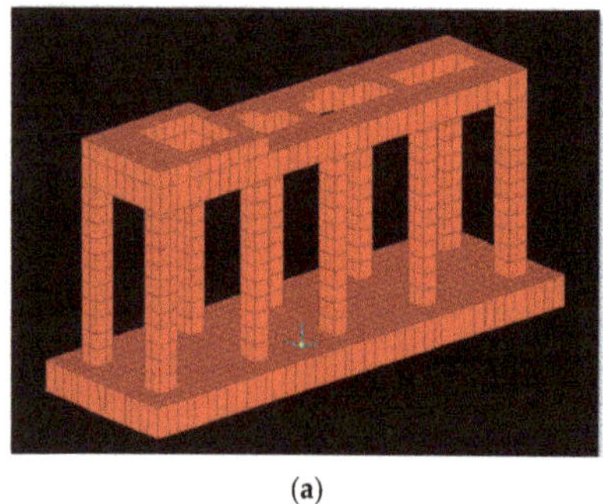
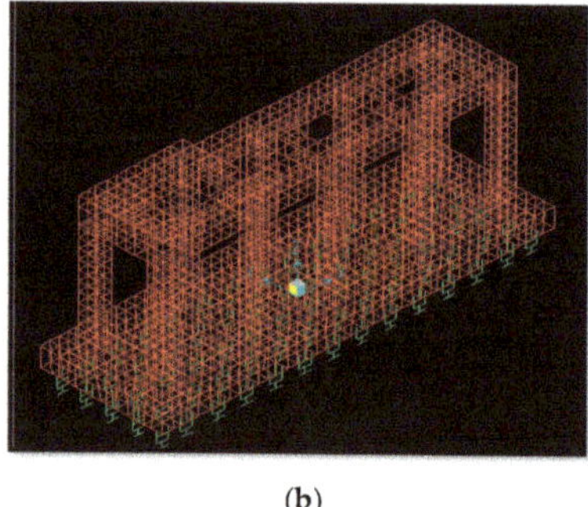

(a)   (b)

Figure 8. 3-D solid element model and FE link element: (**a**) 3-D Solid element model (31 m × 10 m × 12.5 m), (**b**) FE link element (Green).

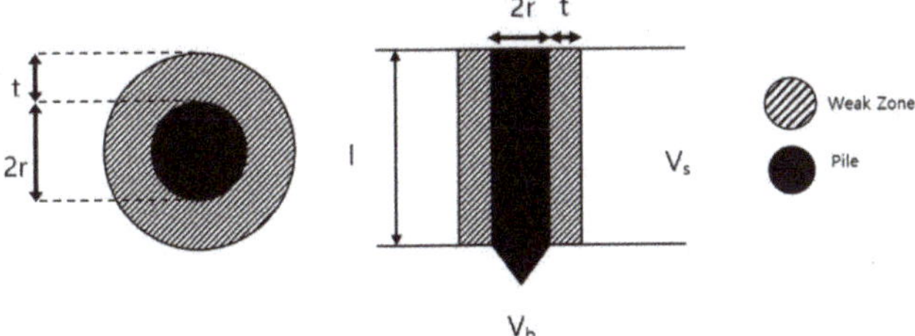

Figure 9. Soil-pile section and weak zone (l: Pile length, r: Radius of the pile, l/r: Slenderness ratio of the pile, t: Thickness of the weak zone, vs.: Shear wave velocity of soil, V_b: Shear wave velocity below the pile tip).

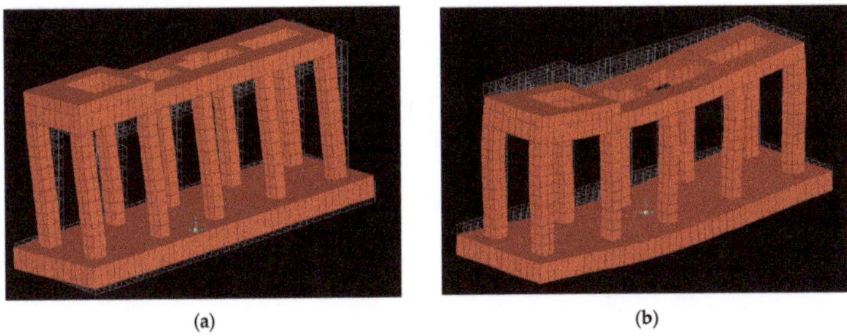

(a)   (b)

Figure 10. Foundation's mode shapes. (**a**) 1st mode 0.395 s (Horizontal mode with rocking), (**b**) 4th mode 0.099 s (vertical mode).

The individual soil-pile interactions were numerically simulated in the DYNA6 platform [8]. The dynamic pile stiffness and damping along the frequency range have been generated and plotted in Figure 11. The stiffness and damping values were incorporated into the FE model (Figure 8, Right) via frequency-dependent link elements in SAP2000.

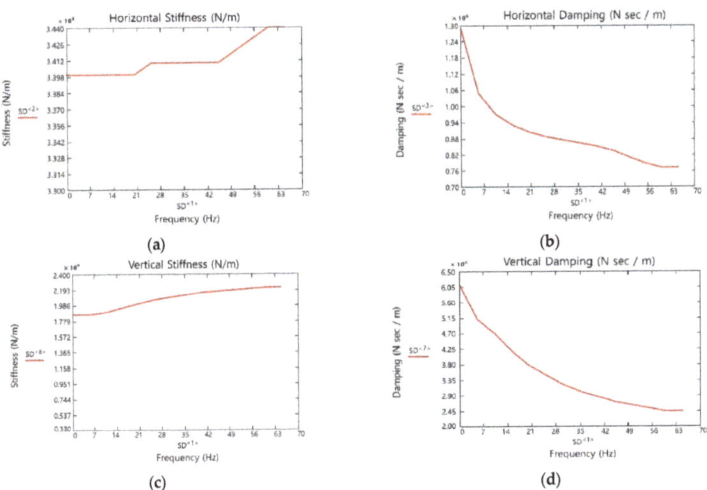

(a)   (b)

(c)   (d)

Figure 11. Soil-pile stiffness and damping vs. frequency: (**a**) Horizontal stiffness vs. frequency, (**b**) Horizontal damping vs. frequency, (**c**) Vertical stiffness vs. frequency, (**d**) Vertical damping vs. frequency.

In general, the horizontal stiffness has little variation with the frequency range (less than 2%) while the vertical pile stiffness gradually increases up to 30%. Horizontal and vertical damping is drastically reduced as frequency increases. In particular, the frequency-dependent values are very different from frequency-independent values, as shown in Table 6. The frequency-dependent vertical stiffness is 10–13 times higher than the frequency-independent value. The frequency-dependent horizontal stiffness is 3.5 times higher.

Table 6. Individual pile stiffness and damping comparison.

Range 0–70 Hz	Vertical Stiffness (N/m)	Vertical Damping (N sec/m)	Horizontal Stiffness (N/m)	Horizontal Damping (N sec/m)
Frequency Independent	1.734×10^8	2.203×10^6	9.89×10^7	1.392×10^6
Frequency Dependent	$1.82 \times 10^9 \sim$ 2.2×10^9	$6.05 \times 10^6 \sim$ 2.45×10^6	$3.398 \times 10^8 \sim$ 3.44×10^8	$1.3 \times 10^6 \sim$ 7.6×10^5
Delta (0–70 Hz)	10~13 Times Increase	3~1 Times Increase	3.4~3.5 Times Increase	0.9~0.5 Times Reduction

8. FE (Finite Element) Steady-State Vibration Analysis

Frequency domain steady-state analysis has been proposed to calculate only the permanent vibration amplitude from the basic equation of motion as follows.

$$[M]\{u''(t)\} + [C]\{u'(t)\} + [K]\{u(t)\} = \{p(t)\} \tag{3}$$

Converting to complex form, then the following is true.

$$[K + iwC - w^2M]\, a = p \tag{4}$$

in which $K = [K + iwC - w^2M]$ is a complex impedance matrix, $K - w^2M$ represents stiffness and the inertia effect, and iwC is the imaginary part considering the damping effect. Thus, the equation of motion can be expressed with a function of frequency as follows.

$$K(w)\, a(w) = p(w) \tag{5}$$

The solution from Equation (5) for multi-degrees-of-freedom (MDOF) can be easily calculated through the finite element analysis solver. For the given vendor's dynamic unbalanced forces (F) at operating speed, one can generate: em factor $= F/(2\pi f)^2$. The steady-state function value can be given by $\omega^2 = (2\pi f)^2$. Taking advantage of the efficient steady-state solver in SAP2000, any nodal amplitude of interest can be generated, along with the frequency range. This gives a full-spectrum view of dynamic amplitudes over the frequency range and elevates the reliability of a vibrational assessment. For instance, several nodal amplitudes of the compressor foundation are plotted in Figure 12 for a hysteretic damping of 1% and 2%. The frequency-domain analysis results show that the dynamic amplitudes reflect a significant and meaningful engineering mode and modal responses in the sense of resonance checking. The hysteretic damping effect on the amplitude calculation over the range is clearly shown. Therefore, the importance of the selection of a reasonable damping value should be emphasized. From these observations, the steady-state vibration analysis technique enables foundation engineers to utilize the exact soil-pile stiffness and damping at certain frequencies for the corresponding amplitude calculations over such ranges.

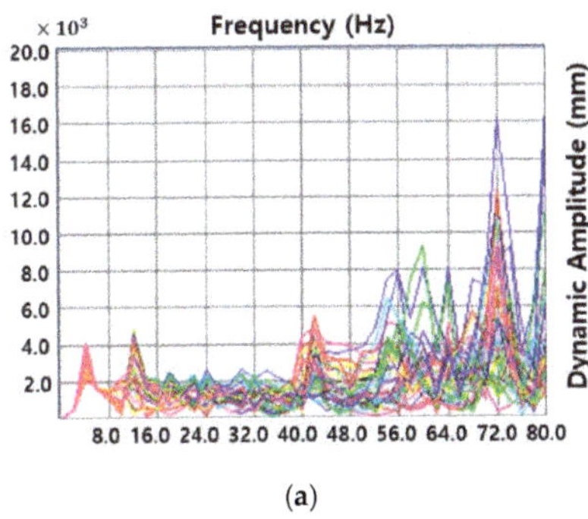
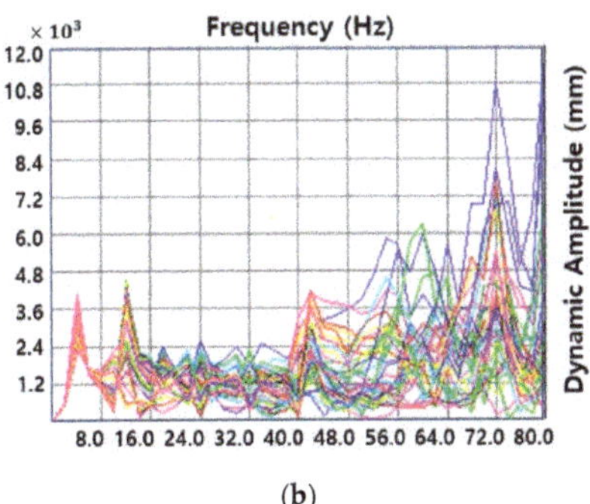

(a)   (b)

Figure 12. Steady-state vibration analysis considering frequency-dependent interaction: (**a**) hysteretic damping = 1% and (**b**) hysteretic damping = 2%

9. Discussion and Conclusions

The steady-state solutions by modal superposition analyses and direct integration time history analyses of the CTG and STG foundations were computed using several combinations of damping options, which would be generally available in commercial FE software. Frequency-domain steady-state analysis using a frequency-dependent soil-pile interaction was developed, and the effect on hysteretic damping was addressed. Based on the vibration response of the subject foundations, the following conclusions and recommendations can be made from this study.

1. The steady-state vibration velocity responses computed by SAP2000 and GTSTRUDL using modal superposition analysis shows good agreement in both CTG and STG foundations. The cases where there were minor differences are due to the consideration of the incompatible bending mode in the SAP2000 model, which significantly improves local bending behaviour at higher modal frequencies.

2. Modal damping analyses using 4% damping produce consistent vibration performances compared to those with soil radiation damping of 20% (cut-off by EPRI) plus 2% material damping. In other words, the total modal damping by the latter method, which largely depends on the deformation of the soil-foundation interface at the significant modal frequency, would be close to 4%. Modal superposition analyses using 2% and 10% modal damping to some extent overestimate and underestimate the vibration level, respectively. The direct use of damping ratios by the Elastic Half-Space Solution, without cut-off limits, results in an underestimation of the vibration response.

3. The two frequency-point based Rayleigh damping (proportional to mass and stiffness) in the direct integration analysis approximates the replacement of the participating modal damping. The CTG foundation's vibrational responses were almost the same, regardless of the variation of the first modal damping from 4% to 20%, together with 4% damping at 60 Hz. This is primarily due to the fast drop of damping after the first mode and the fast convergence of damping toward 4% until 60 Hz. Furthermore, the 60 Hz vibration response is generally less affected by the first modal damping. After the 60 Hz threshold, the Rayleigh damping increases gradually, but the increase would be negligible for the 60 Hz response.

4. Frequency-dependent vertical and horizontal stiffness gradually increased as the frequency grew. However, frequency-dependent vertical and horizontal damping decreased rapidly as the frequency increased.

5. A full-spectrum steady-state vibration solution compensates for the disadvantage of the modal time history solution, and, thus, increases the reliability of the foundational design. Foundation engineers will be able to estimate the prospective vibration level in a broad frequency range.

Author Contributions: Conceptualisation, S.-H.S. Methodology, S.-H.S. Validation, S.S.L. Formal analysis, S.-H.S. Data collection, S.-H.S. Writing—original draft preparation, S.-H.S. Writing—review, S.S.L. Supervision, S.S.L. Project administration, S.S.L. Funding acquisition, S.S.L.

Funding: The research fund of Hanyang University (HY-201700000002431) supported this research and the authors greatly appreciate the financial support.

Conflicts of Interest: The authors declare no conflict of interest. The funders had no role in the design of the study, in the collection, analyses, or interpretation of data, in the writing of the manuscript, or in the decision to publish the results.

References

1. Novak, M. Dynamic Stiffness and Damping of Piles. *Can. Geotech. J.* **1974**, *11*, 574–598. [CrossRef]
2. Kausel, E.; Ushijima, R. *Vertical and Torsional Stiffness of Cylindrical Footing*; Civil Engineering Department Report R79-6; MIT: Cambridge, MA, USA, 1979.
3. Roesset, J.M. Stiffness and Damping Coefficients of Foundations. In Proceedings of the APEC Session on Dynamic Response of Pile Foundation: Analytical Aspects, New York, NY, USA, 30 October 1980; pp. 1–30.
4. Arya, S.C.; O'Neill, M.W.; Pincus, G. *Design of Structures and Foundations for Vibrating Machines*; Gulf Publishing Company: Houston, TX, USA, 1979; pp. 1–90.
5. Electric Power Research Institute (EPRI). *Fan Foundation System—Analysis and Design Guidelines, Project 1649-3 Final Report*; CS-4746; EPRI: Palo Atlo, CA, USA, 1986; pp. 1–5.
6. Novak, M.; Aboul-Ella, F. Stiffness and Damping of Piles in Layered Media. In Proceedings of the Earthquake Engineering and Soil Dynamics, ASCE Specialty Conference, Pasadena, CA, USA, 19–21 June 1978; pp. 704–819.
7. Novak, M.; Aboul-Ella, F. Impedance Functions of Piles in Layered Media. *J. Eng. Mech.* **1978**, *104*, 643–661.
8. *DYNA6 User Manual*; Dynamic Analysis of Foundations for the Effects of Harmonic, Transient and Impact Loadings; Geotechnical Research Center, University of Western Ontario: London, ON, Canada, 2011.
9. *SAP2000 CSI Analysis Reference Manual*; Computers and Structures, Inc.: Berkeley, CA, USA, 2007.
10. *GT STRUDL Analysis User Guide*; Hexagon: North Kingstown, RI, USA, 2017.
11. Chopra, A.K. *Dynamics of Structures, Theory and Applications to Earthquake Engineering*; Prentice Hall: Englewood Cliffs, NJ, USA, 1995; pp. 35–60.
12. Chowdhury, I.; Dasgupta, S.P. Computation of Rayleigh Damping Coefficients for Large Systems. *Electron. J. Geotech. Eng.* **2003**, *8*, 1–11.

Article

Seismic Behavior of Steel Plate-Concrete Shear Walls with Holes

Zhihua Chen [1,2], Jingshu Wu [2,3,*] and Hongbo Liu [1,2]

[1]   Department of Civil Engineering, Tianjin University, Tianjin 300072, China; zhchen@tju.edu.cn (Z.C.); hbliu@tju.edu.cn (H.L.)
[2]   State Key Laboratory of Hydraulic Engineering Simulation and Safety, Tianjin 300072, China
[3]   Research Institute of Building and Construction CO.LTD.MCC Group, Beijing 100088, China
*   Correspondence: wujingshu@nqstc.com

Received: 11 October 2019; Accepted: 27 November 2019; Published: 3 December 2019

Abstract: Steel plate-concrete shear walls (SPSW) are used as the containment in nuclear power stations. However, the influence of holes and axial loading on the behavior of steel plate-concrete shear walls is neglected in most studies. Thus, it is necessary to understand the seismic behavior of SPSW members with holes in order to avoid the potential risks for nuclear power stations. In this study, a series of specimens were tested by low-cycle reciprocal loading. The specimens were designed with different holes to simulate real members in nuclear power stations. A hysteretic curve of specimens was obtained from a low-cycle reciprocal test to discuss the seismic behavior of steel plate-concrete shear walls (SPSW). Moreover, effects of axial compression ratio, hole size, thickness of the steel plate, and hole position on the hysteretic performance of SPSW were analyzed. The horizontal ultimate bearing capacity of SPSW specimens was estimated using the norms of the Architecture Institute of Japan and the calculation method of Ono reduction rate. Results provide theoretical references for the design and application of SPSW with holes.

Keywords: SPSW; axial compression ratio; hole size; thickness of stiffening plate; hole position; hysteretic performance

---

1. Introduction

Steel plate-concrete shear walls (SPSW) serve as the containment of AP1000 and CAP1400 nuclear power stations and of the stress components of the internal plants in nuclear power stations. Structures with SPSW have advantages of modular construction and strong seismic and impact resistance. In the 1960s, Japan began to apply the steel stiffening concrete seismic structural structure. A series of studies on shear capacity and the stiffness and ductility of SPSW structures have been conducted in many countries, such as the US and Japan [1]. Other relevant research is based on low-cycle reciprocal tests. Lubell et al. [2] conducted an experimental testing on two single and one four-story steel shear wall specimens under cyclic quasi-static loading. Hjjar et al. [3] performed a low-cycle reciprocal test on a 1:2 scale model of SPSW structure and found its high bearing capacity, energy-consumption mechanism, and good ductility. The test results show that the superposition principle is basically true [4]. Berman et al. [5] conducted an experiment on the light-gauge steel plate shear walls and braced frames to study the hysteretic behavior. They revealed that the energy dissipated per cycle and the cumulative energy dissipation are similar for the two structures. Later, Gan, Zhao, and Wang et al. [6–8] studied the seismic behavior of the steel plate-reinforced concrete shear wall by using the quasi-static test. Li and li [9] investigated the out-of-plane seismic behavior of steel plate and concrete infill composite shear walls (SCW). They found that SCW has a better ultimate capacity and lateral stiffness. Huang et al. [10] proposed an innovative concrete-filled double-skin steel plate SCW and investigated its seismic behavior. By conducting a quasi-static cyclic test, the wall is confirmed to have good seismic performance.

In recent years, various scholars have simulated the hysteretic curve and stiffness reduction rate of SPSW through finite element simulation. Rafiei et al. [11] presented and verified the finite element model to simulate the behavior of a novel SCW consisting of the two skins of profiled steel sheeting and an infill of concrete under in-plane loadings. Hu et al. [12] analyzed the moment-curvature behavior of concrete-filled steel plate SCW using refined material constitutive models. Peter et al. [13] presented the development and benchmarking of a detailed 3D nonlinear inelastic finite element model to predict the lateral load-deformation response and behavior of the 1/6th scale test structure. Nguyen et al. [14] presented a numerical study of steel-plate concrete composite walls by using the general-purpose finite element program ABAQUS. The influence of key design variables, including the reinforcement ratio, connector type, and faceplate slenderness ratio, were likewise studied. Wang et al. [15] investigated the hysteretic performance of the SPSW wall by using Open Sees software. Moreover, parameters such as the steel plate ratio, axial compressive load ratio, concrete strength, and web reinforcement ratio were analyzed. Yamatani [16] performed a low-cycle reciprocal test of SPSW with holes at the lower position under the shear-span ratio of 0.7, a distance–thickness ratio of 100, and opening area ratio of 0.3. Other scholars proposed the concept of reduction coefficient to evaluate the bearing capacity of an SPSW wall. Satou Kouichiet al. [17] conducted a numerical simulation of an SPSW structure and found that a numerical simulation is applicable for the calculation and analysis of the performance of shear walls with holes. Ishida Masatoshi [18] analyzed the seismic behavior of an SPSW structure by using theories and finite element simulation. Fujita Tomohiro and Oosuka et al. [19,20] performed anti-shear tests on SPSW structures with holes. Adding sleeves and using increased thickness surface steel plates on the shear wall were determined to be effective reinforcement methods. Some scholars also used the XFEM (the eXtended Finite Element), XIGA (the eXtended IsoGeometric Analysis) and Jaya algorithm to predict the occurrence of cracks and other defects in walls and slab [21–23].

So far, steel plate-concrete shear walls are studied widely. However, the influence of holes and axial loading on the behavior of steel plate-concrete shear walls are neglected in most studies. Thus, the seismic behavior of steel plate-concrete shear walls is completely different when the influences are considered. In the study, a series of low-cycle reciprocal loading tests are conducted on an SPSW structure with holes, thus obtaining the ultimate bearing capacity and failure mode of the structure. The influences of holes and reinforcing measures and axial loads of components on the seismic behavior of an SPSW structure are analyzed, thus determining the difference between theoretical and test values. The seismic behavior and stress mechanism of SPSW specimens are discussed theoretically.

2. Experiment

2.1. Experimental Apparatus

A low-cycle reciprocal test of an SPSW structure with different hole sizes was conducted in the Beijing Key Laboratory of Engineering Anti-earthquake and Structural Diagnosis of the Beijing University of Technology. The test applied a horizontal servo actuator (maximum range = 2000 kN), which was fixed on the concrete counterforce wall. Test devices included a counterforce wall, counterforce frame, loading beam, bottom beam, and displacement meter. The bottom beam was strongly connected to the ground through a fixing device, and a steel ingot was used on the loading beam. On the one hand, this design prevents concrete crushing caused by excessive force. On the other hand, the vertical load is ensured to be a uniform load. In the test, the low-cycle reciprocal loading test device is composed of horizontal and vertical jacks. In this device, the jack on the transverse counterforce frame applies the vertical loads. During the whole test, these vertical loads are kept as stable as possible through the manual control of the oil pump. The horizontal jack applies the horizontal loads. One end of the jack is connected to the loading beam of the specimens, and the other end is fixed on the counterforce wall (Figure 1).

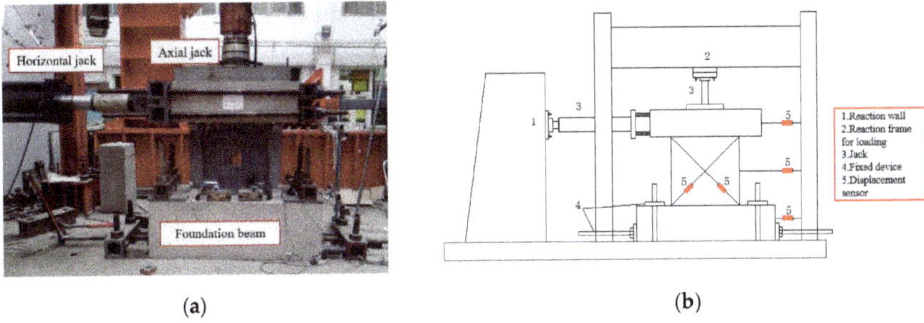

Figure 1. Loading device (**a**) and measurement arrangement (**b**).

2.2. Design of Specimens and Instrumentation

To accurately reflect shielding plants and internal SPSW structures in a nuclear power station, specimens in this study were designed with reference to Japan's *Technical Regulations on Earthquake-resistant Design of Steel Plate Concrete Structure* (JEAC4618-2009) [24] and design documents of AP1000 demonstration projects. According to *Technical Regulations on Earthquake-resistant Design of Steel Plate Concrete Structure*, the ratio between the thickness of SPSW for nuclear safety and thickness of the surface steel plate should be controlled within 30:200. On this basis, thicknesses of the shear wall and surface steel plate were set. Considering the actual welding ability in the laboratory, the thickness of the surface steel plate in the test specimens was determined to be 2.5 mm.

In this study, low-cycle reciprocal tests of nine SPSW specimens were conducted. Dimensions of SPSW specimens were 800 mm (height) × 800 mm (width) × 125 mm (thickness). SPSW specimens were divided into three types: (1) without holes, (2) with small holes, and (3) with large holes. Considering the loading capacity of the laboratory, the SPSW specimens without holes and with small holes were flat (Figure 2), whereas the SPSW specimens with large holes were I-shaped (Figure 3). The stiffening plates were spread around the hole, with a width is 90 mm, and the length determined by the perimeter of the hole. The hole size, thickness of the stiffening plate, hole position, and axial pressure ratio are listed in Table 1. In the SPSW specimens, the steel plate was made of Q235B, and the grade of the concrete strength was C35. Material properties are shown in Tables 2 and 3. Fine aggregates with a diameter smaller than 10 mm were used.

Table 1. Parameters of specimens.

Specimen	Flange Size (mm)		Thickness of Steel Plate (mm)	Hole Size (mm)	Hole Position	Thickness of Stiffening Plate	Axial Compression Ratio	Wall Type
	Width	Thickness						
SCW-1	–	–	2.5	–	–	–	0.3	Flat
SCW-2	–	–	2.5	125 × 125	Center	2.5	0.15	Flat
SCW-3	–	–	2.5	125 × 125	Center	2.5	0.3	Flat
SCW-4	–	–	2.5	125 × 125	Center	2.5	0.5	Flat
SCW-5	–	–	2.5	125 × 125	Center	3.5	0.3	Flat
SCW-6	–	–	2.5	125 × 125	Center	2.5	0.15	Flat
SCW-7	–	–	2.5	180 × 180	Center	2.5	0.15	Flat
SCW-8	395	110	2.5	430 × 600	Center	2.5	0.15	I-shaped
SCW-9	395	110	2.5	430 × 600	Eccentricity	2.5	0.15	I-shaped

Table 2. Mechanical properties of the steel plate.

Thickness of Steel Plate (mm)	Yield Strength (MPa)	Tensile Strength (MPa)	Elasticity Modulus (MPa)	Maximum Elongation (%)
2.5	369	486	2.02×10^5	28.5

Table 3. Mechanical properties of the concrete.

Material Number	Axial Compressive Strength (MPa)	Axial Tension Strength (MPa)	Elasticity Modulus (MPa)
C35	25.98	2.33	3.25×10^4

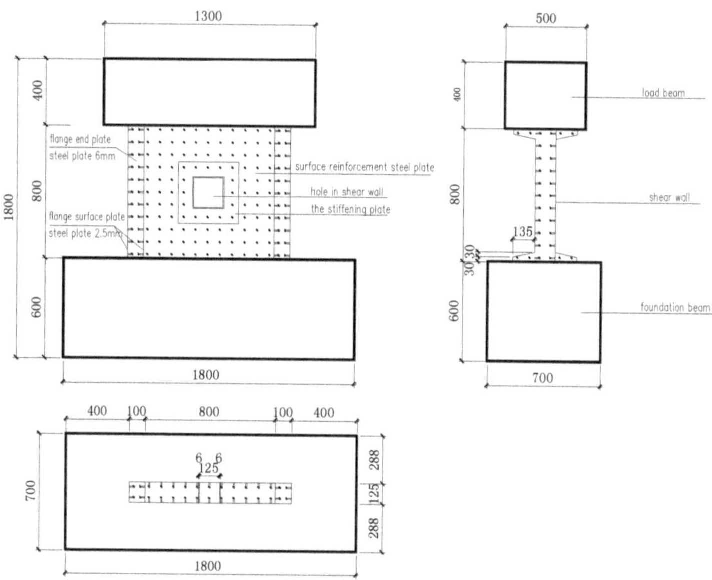

Figure 2. Typical flat specimens.

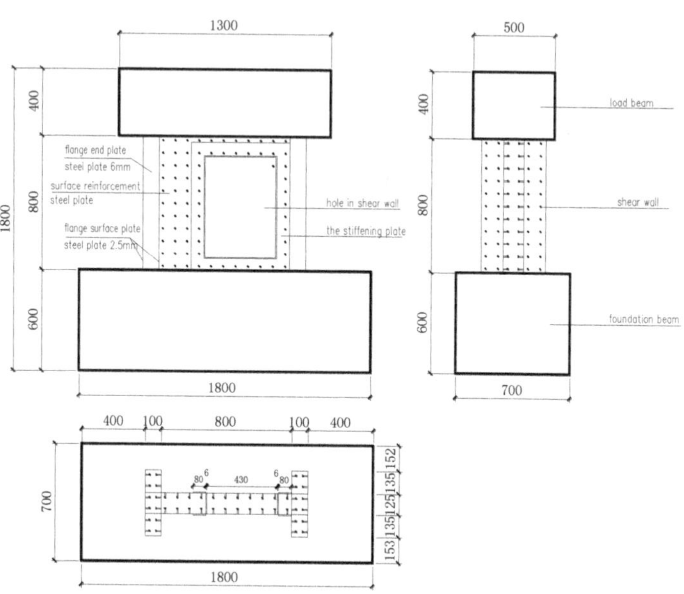

Figure 3. Typical I-shaped specimens.

In the test, five displacement sensors were set (Figure 1), three of which were set vertically on the wall. The first displacement sensor on the side of the loading beam of the shear wall tested the top wall displacement under reciprocal loads. The displacement sensor in the center of the shear wall examined the bottom slippage during the loading process. The displacement sensor on the side of the bottom beam tested the overall slippage of specimens to assure test accuracy. In addition, two dial indicators were installed at diagonal positions of the wall to test for outward deflection.

2.3. Loading Mode

The loading process is composed of pre-loading and formal loading. First, the vertical jack applied the vertical loads, and then 20kN force was pre-loaded by a horizontal actuator. Second, loads were removed to ensure contact of the loading device with the specimens. The loading protocol refer to construction standards of the China Construction Industry JGJ/T 101-2015 [25].

In the formal loading, the vertical jack also applied the vertical loads on the specimens. Before full loading, the vertical loads were applied 2–3 times according to the design value of 40–60% to eliminate the influences of the internal non-uniformity of shear wall specimens. When the vertical load was applied to the designated value and stabilized, a horizontal actuator was used to apply the horizontal low-cycle reciprocal loads to the specimens. Load control was applied in the early stage of the test. Loads in the first, second, third, and fourth cycles were 50, 100, 200, and 400 kN, respectively. Later, loads were increased by 100 kN every cycle until specimens fail. The loading system is illustrated in Figure 4. Specimens were considered a failure upon attaining one of the following cases:

(1)   The bearing capacity of specimens decreased to lower than 85%.
(2)   Serious failure occurred in specimens, such as heaving of steel plates or crushing of concrete, thereby resulting in a difficult loading.

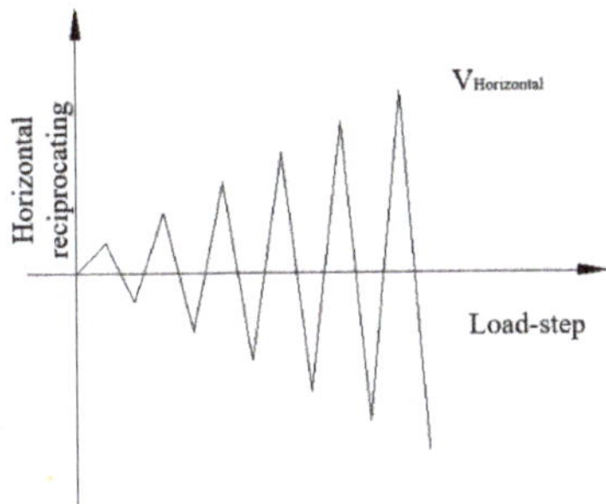

Figure 4. Loading system.

3. Experimental Analysis

3.1. Failure Mode

Specimen SCW-1 is an SPSW without holes. The loading process of SCW-1 was relatively stable. When the cyclic loading approaches 700 kN, the concrete begins to crush, and the steel plate develops slight deformation. With the continuous increase in loads, the crushing region in the concrete and deformation of steel materials likewise continued. Heaves develop on the steel plate at the root side of the shear wall until the load reaches 1100 kN. When the cyclic loads increase to 1240 kN, heaves at the root side of the steel plate of the shear wall and surface steel plate intensify, accompanied with the crushing of concrete at root positions. Due to the processing deviation of specimens and influences of initial defects, deformations of steel plates at the two root sides of the shear wall are inconsistent but are symmetric to an extent.

In the tests of flat specimens with small holes, the horizontal loads are small in the initial loading stage, and specimens are still elastic. A linear relationship is observed between the horizontal load and vertex displacement of specimens. In the late stage, internal concrete begins to crack as horizontal loads increase. The side and surface plates yield successively. As lateral displacement increases, the steel plate begins to bend or crack. The weld joints of several specimens are pulled open, and internal concretes are crushed, resulting in failure (Figure 5). The main failure develops at the wall bottom and is evaluated as brittle failure. Several specimens develop slippage due to the lower stiffness of the bottom beams.

(**a**) buckling of side plates and pulling failure of weld joints   (**b**) concrete crushing

Figure 5. Bucklingof side plates, pulling failure of weld joints, and concrete crushing of flat steel plate-concrete shear wall (SPSW) specimens with small holes.

Figure 6 shows the failure mode of I-shaped SPSW specimens with large holes. Bending shear failure is the dominant failure mode. In the test of SCW-8, the internal concrete in SPSW begins to crack at approximately 500 kN. Furthermore, the steel plate at the holes begins to yield and pull open at approximately 800 kN. The surface steel plate yields at 1100 kN. Subsequently, the surface plate cracks and bends, accompanied with wrinkles. The internal concrete of the steel plate is crushed after losing constraints. Moreover, the steel plate at the wing wall side buckles, and the wall develops serious deformation. The bearing capacity of SPSW specimens continue to decrease, and the loading test is terminated. According to the results, SCW-8 with holes at the center and SCW-9 with holes biased to the axis have no significant differences in terms of bearing capacity and failure mode.

(**a**) top left of SCW-8   (**b**) bottom right of SCW-8

Figure 6. *Cont.*

(c)failure mode of SCW-8   (d)failure mode of SCW-9

Figure 6. Failuremodes of I-shaped SPSW specimens with large holes.

3.2. Effects of Axial Compression Ratio on the Hysteretic Performance of SPSW Specimens

The axial compression ratio is an important factor that influences the hysteretic performance of SPSW specimens. In this section, axial compression ratios in the low-cycle reciprocal tests of three SPSW specimens with holes (SCW-3, SCW-4, and SCW-5) are set at 0.15, 0.3, and 0.5, respectively. The ultimate loads and displacements of specimens are shown in Table 4. Figure 7 illustrates the hysteretic curves of different SPSW specimens, whereas Figure 8 shows the skeleton curves.

Table 4. Ultimate loads and displacements of specimens.

Specimens	Ultimate Loads (kN)		Ultimate Displacements (mm)	
	Positive	Negative	Positive	Negative
SCW-3	1100	−1074	18.09	−27.05
SCW-4	1217	−1185	18.46	−28.60
SCW-5	1362	−1301	16.56	−13.88

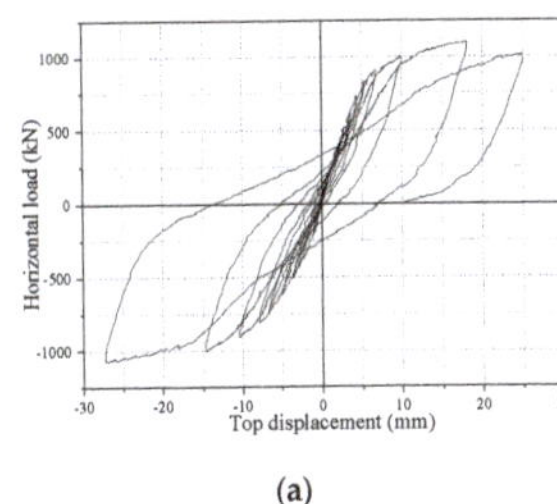

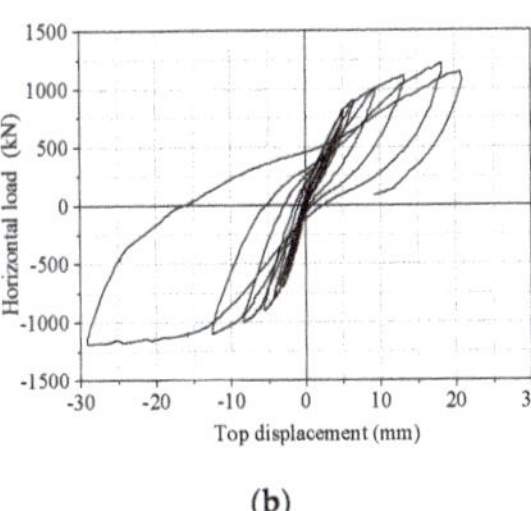

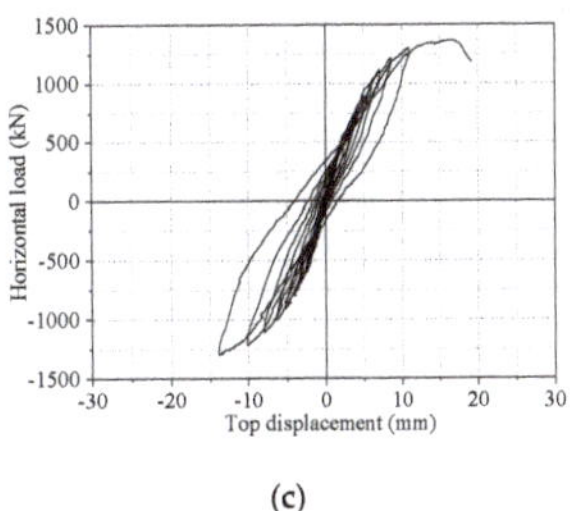

(a)   (b)   (c)

Figure 7. Hysteretic curve of specimens under different stress ratios. (**a**) SCW-3, (**b**) SCW-4, and (**c**) SCW-5.

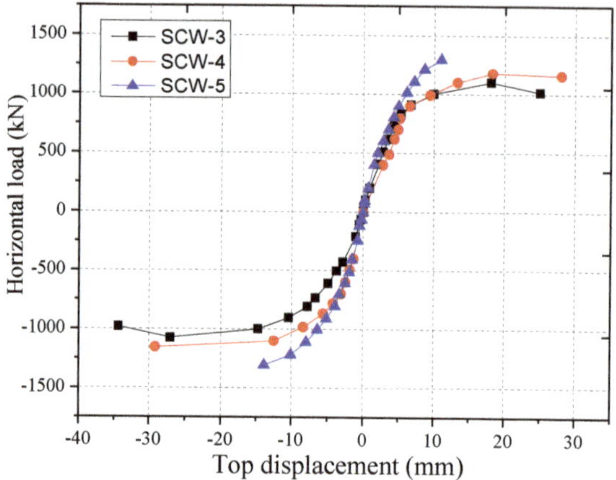

Figure 8. Comparison of the skeleton curves of specimens of the same size under different axial compression ratios.

In the loading process of the three specimens, steel plates develop low sounds, and internal concretes begin to crack at 350 kN when the axial compression ratio is 0.5. When the axial compression ratios are 0.15 and 0.3, steel plates develop low sounds at 500 kN and 150 kN, respectively. In summary, the concrete begins to crack early under large axial compression ratios. With the continuous increase of loads, SCW-3 and SCW-4 show a reduction stage of loads and good ductility. When loads increased to 1217 kN, the root plate of SCW-4 bent, and the test was terminated. SCW-4 thus shows poor ductility, as verified from the skeleton curve. Table 4 shows that the ultimate bearing capacity of components is positively related with axial compression ratio, but deformation resistance and ductility are negatively correlated. Therefore, attention should be paid to the sudden failure of structures under high axial compression ratio.

3.3. Effects of Hole Area on the Hysteretic Performance of SPSW Specimens

Influences of hole size on lateral bearing capacity, deformation resistance, and energy consumption of SPSW specimens with different hole sizes are determined through a low-cycle reciprocal test. Ultimate strengths and displacements of SPSW specimens with different hole sizes are presented in Table 5. Figures 9 and 10 respectively display the hysteretic and skeleton curves of SPSW specimens with different hole sizes.

Table 5. Ultimate loads and displacements of specimens.

Specimens	Hole Size (mm)	Ultimate Loads (kN)		Ultimate Displacements (mm)	
		Positive	Negative	Positive	Negative
SCW-1	–	1277	−1310	23.15	−19.27
SCW-3	125 × 125	1100	−1074	18.09	−27.05
SCW-7	180 × 180	1190	−1105	37.43	−27.45
SCW-8	430 × 600	1055	−1007	51.94	−43.49

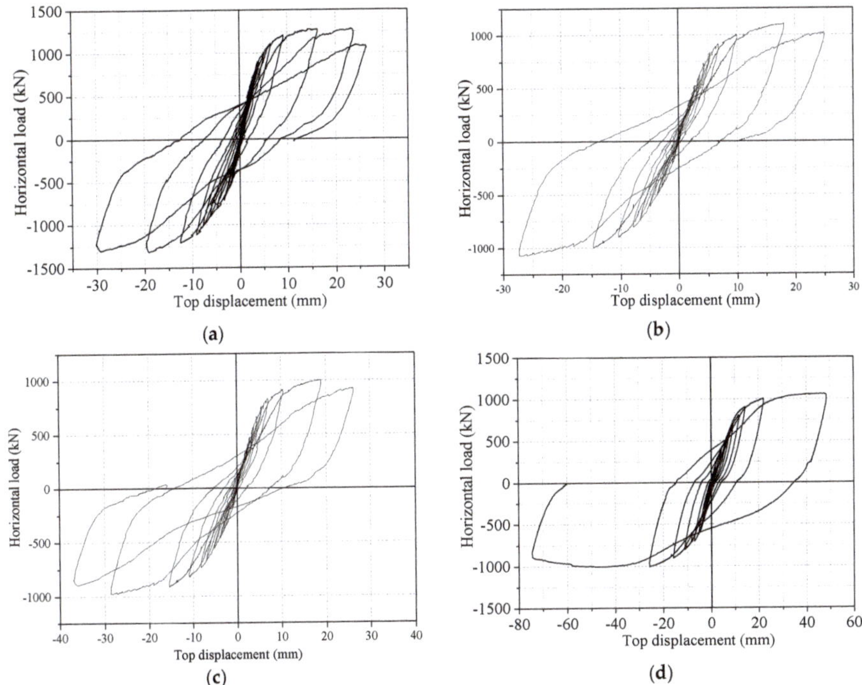

Figure 9. displacement curves of SPSW specimens with different hole sizes. (**a**) SCW-1, (**b**) SCW-3, (**c**) SCW-7, and (**d**) SCW-8.

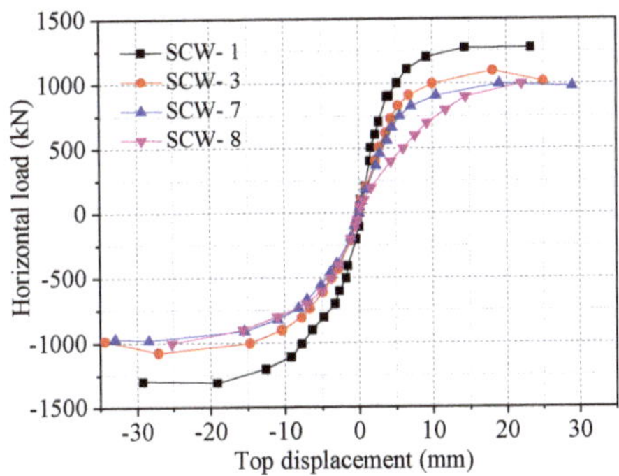

Figure 10. Skeleton curves of SPSW specimens with different hole sizes.

Table 5 shows that as hole area increases, the ultimate displacement of specimens gradually increases, but the ultimate loads of specimens decrease slightly. Hysteretic curves of SCW-1, SCW-3, and SCW-7 have similar shapes. They are full but have clear twist contraction effects. Compared with the first three specimens, SCW-8 has a fuller hysteretic curve, indicating better seismic behavior.

3.4. Effects of the Thickness of the Stiffening Plate on the Seismic Behavior of SPSW Specimens

SCW-1, SCW-4, and SCW-6 have the same appearance size. SCW-1 has no hole, SCW-4 and SCW-6 have the same geometric size, hole position, and hole size but are equipped with 2.5 mm and 3.5 mm stiffening plates, respectively. Table 6 shows that compared with the uncut SCW-1, the strength and stiffness of the open specimens are decreased. Compared with the ultimate strength of SCW-4, the ultimate strength of SCW-6 increased by 4% in the positive direction and 10% in the negative direction. The hysteretic curves of SCW-4 and SCW-6 are shown in Figure 11, and the skeleton curves of specimens are shown in Figure 12. The hysteretic curve of SCW-6 is S-shaped, which is attributed to slippage caused by anchoring failure. Similarly, the SCW-6 skeleton curve displays that its initial stiffness is low due to the large specimen displacement.

Table 6. Ultimate loads and displacements of specimens.

Specimens	Ultimate Loads (kN)		Ultimate Displacements (mm)	
	Positive	Negative	Positive	Negative
SCW-1	1277	−1310	23.15	−19.27
SCW-4	1217	−1185	18.46	−28.60
SCW-6	1260	−1302	21.41	−28.11

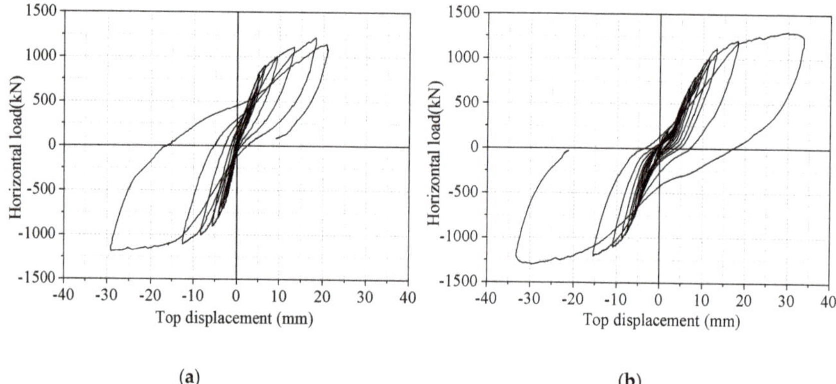

(a)   (b)

Figure 11. Load displacement curve of specimens reinforced by different thicknesses of plates. (**a**) SCW-4 and (**b**) SCW-6.

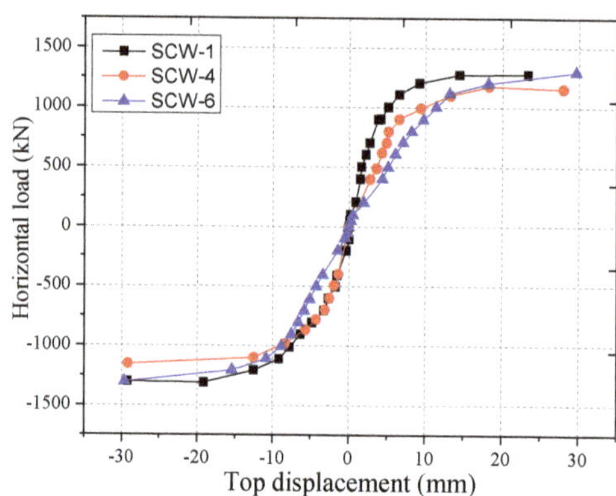

Figure 12. Skeleton curves of specimens reinforced by different thicknesses of plates.

3.5. Effects of Hole Position on the Seismic Behavior of SPSW Specimens

Figures 13 and 14 illustrate the hysteretic and skeleton curves of specimens with large holes at different positions as determined through low-cycle reciprocal tests, respectively. The ultimate strengths and displacements of specimens are presented in Table 7.

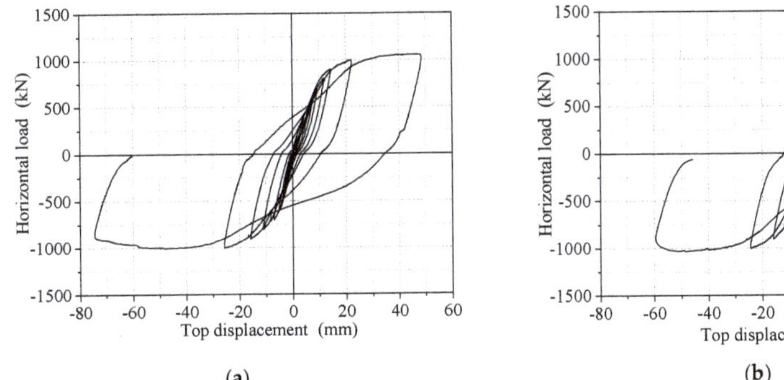

(a)   (b)

Figure 13. Hysteretic curves of specimens with holes at different positions. (**a**) SCW-8 and (**b**) SCW-9.

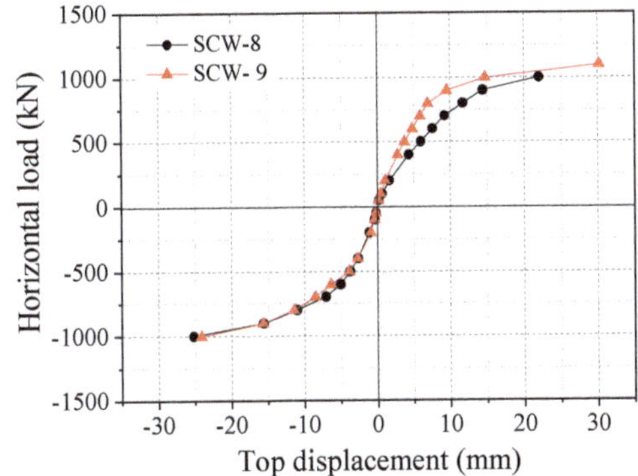

Figure 14. Skeleton curves of specimens with holes at different positions.

Table 7. Ultimate loads and displacements of specimens.

Specimens	Ultimate Loads (kN)		Ultimate Displacements (mm)	
	Positive	Negative	Positive	Negative
SCW-8	1055	−1007	51.94	−43.49
SCW-9	1103	−1031	34.14	−44.41

Table 7 shows that the ultimate displacement of the specimen with a hole in the center is higher than that of the specimen with an eccentric hole. Comparing the two specimens, the hysteretic curve of the specimen with a hole in the center is fuller, and its energy-dissipation capacity is larger. Hence, placing the hole in the center of SPSW structures is suggested for practical engineering.

4. Theoretical Calculation

In order to estimate the ultimate bearing capacity and lateral stiffness resistance of SPSW specimens with holes, two calculation methods are discussed. The first method is the reduction rate calculation method of the Architecture Institute of Japan (AIJ) [26], and the second method is the Ono method [27]. A comparison between the experimental results and the theoretical results are discussed.

4.1. ReductionRate Calculation Method of AIJ

The calculation methods of the reduction rates of horizontal bearing capacity (r_u) and stiffness (r_c) of shear wall components with holes are regulated in the structural design codes of AIJ. These two calculation formulas are as follows:

$$r_u = 1 - \eta, \tag{1}$$

$$r_c = 1 - 1.25\sqrt{\frac{h_0 l_0}{hl}}, \tag{2}$$

where h_0 and l_0 are the height and width of the hole, h is the floor height, l is the center distance of frame columns, η is the hole area ratio with value $\eta = \max\left\{\sqrt{\frac{h_0 l_0}{hl}}\right\}$, and where $\sqrt{\frac{h_0 l_0}{hl}} \leq 0.4$. In AIJ codes, the application of these formulas is regulated; they are applicable to situations when the hole area is smaller than 0.4 but not for situations when the hole area is larger than 0.4. In these formulas, the shape and position of holes are not considered when calculating bearing capacity and stiffness. That is, the calculated results of bearing capacity and stiffness are the same for specimens with the same hole size.

4.2. Ono Reduction Rate Calculation

Ono Masasyuki, a Japanese scholar, proposed the calculation method of Ono reduction rate through a series of experiments wherein the influences of hole positions are considered [27] (Figure 15).

$$r_u = \sqrt{\sum A_{ei}/hl}, \tag{3}$$

when $0.1 \leq \gamma \leq 0.53$,

$$r_e = \left(\frac{0.025}{0.0303\gamma}\right) \times \alpha\beta + \left(\frac{0.6}{0.55\gamma}\right)/\left(\frac{1.55}{\gamma^2}\right)^k, \tag{4}$$

when $0.53 \leq \gamma \leq 1.00$,

$$r_e = \left(\frac{5.24}{717.24\gamma}\right) \times \alpha\beta + \left(\frac{0.6}{0.55\gamma}\right)/\left(\frac{1.55}{\gamma^2}\right)^k, \tag{5}$$

where A_{ei} is the area of shadow part, h is the floor height, and l is the center distance of frame columns. $\alpha\beta = 2bD/tl'$; where bD is the sectional area of columns, t is the wall thickness, and l' is the net span. $\gamma = 2l_w/l'$; and where l_w is the length of the wall after the hole is removed. $\kappa = h_0/h'$, where h_0 is the height of the hole and h' is the net height of the wall.

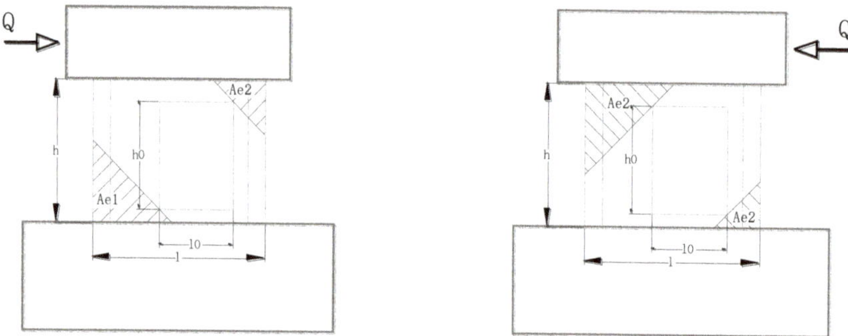

Figure 15. Effective supporting area in the Ono reduction rate calculation method.

In this study, the horizontal ultimate bearing capacity of SPSW members with holes can be obtained by multiplying the reduction rate by that of the corresponding members without holes. The lateral stiffness of SPSW members with holes can be obtained with the same method. Change of the concrete compression zone is considered in the method of Ono reduction rates. The reduction rates are different in different directions for the specimen with biased holes.

Table 8 displays the comparison between the test results of horizontal ultimate loads and the calculated results based on strength reduction rate. Table 9 compares the test results of lateral stiffness and calculated results based on the stiffness reduction rate. This test has no effective lateral stiffness of I-shaped SPSW structures without holes, and only the comparisons between the calculated and test results of SCW-4 and SCW-6 are presented. According to the average ratio, the calculated value of the Ono formula is closer to the experimental value than that of the code formula.

Table 8. Comparison of ultimate loads.

Specimen	Stress State	Q_{test} /kN	Q_{code} /kN	Q_{code}/Q_{test}	Q_{ono} /kN	Q_{ono}/Q_{test}	Axial Compression/kN	Type of Walls
SCW-4	Stressed	1185	1127	0.95	1135	0.96	950	Flat
	Tensioned	1217	1098	0.90	1106	0.91		Flat
SCW-6	Stressed	1302	1127	0.87	1269	0.98	950	Flat
	Tensioned	1260	1098	0.87	1237	0.98		Flat
SCW-8	Stressed	1007	734	0.73	820	0.81	320	I-shaped
	Tensioned	1055	734	0.70	857	0.81		I-shaped
SCW-9	Stressed	1031	734	0.71	836	0.81	320	I-shaped
	Tensioned	1103	734	0.67	836	0.76		I-shaped
Average ratio				0.80		0.88		

Notes: Q_{test} is the test result of horizontal ultimate loads. Q_{code} and Q_{ono} are horizontal ultimate bearing capacities calculated by AIJ codes and the Ono calculation formula.

Table 9. Comparison of lateral stiffness.

Specimens		K_{test}	K_{code}	K_{code}/K_{test}	K_{ono}	K_{ono}/K_{test}	Axial Compression/kN
SCW-4	Negative	41.43	56.11	1.35	67.64	1.63	950
	Positive	65.90	45.51	0.69	54.87	0.83	
SCW-6	Negative	46.31	56.11	1.21	67.64	1.46	950
	Positive	58.86	45.51	0.77	54.87	0.93	

Notes: K_{test} is the test result of lateral stiffness. K_{code} and K_{ono} are stiffness calculated by AIJ codes and the Ono calculation formula.

Table 9 shows that the calculated results of AIJ codes are relatively safe. Moreover, the difference of bearing capacity along different loading directions under eccentric holes is neglected, thus resulting in the low accuracy. The Ono calculation formula considers the influences of loading direction and reflects the effects of hole position on horizontal bearing capacity along different loading directions. The calculation accuracy of the Ono calculation formula is thus higher than that of AIJ codes with respect to specimens with small holes (SCW-4 and SCW-6) and with large holes (SCW-8 and SCW-9). With respect to stiffness, the bottom slippage of the wall can influence the test results to a certain extent, resulting in the difference in stiffness along the two loading directions. Therefore, the calculation method may differ. In summary, the calculation formulas of AIJ codes and Ono are feasible in determining the influences of holes on the horizontal bearing capacity of SPSW structures. Compared with the calculation results of AIJ codes, those of the Ono formula are closer to the actual test results, indicating its high calculation accuracy. Due to the bottom slippage of walls, the rigidity in the negative direction is lower than the calculated values. In the positive direction, the calculated values are lower than the test results, which is due to the fact that the AIJ code and Ono formula are used for structure design and the result are relatively safer.

5. Conclusions

Steel plate-concrete shear walls are studied widely. However, the influence of holes and axial loading on the behavior of steel plate-concrete shear walls are neglected in most studies. Thus, the seismic behavior of steel plate-concrete shear walls is completely different when the influences are considered. Thus, a series of low-cycle reciprocal loading tests were conducted on SPSW specimens with holes in this study. The failure modes and hysteretic curves of SPSW structures under different working conditions were also studied. On the basis of failure modes and hysteretic curves, the seismic behavior of SPSW structures was discussed. Influences of axial compression ratio, hole size, hole position, and the thickness of stiffening plate on the seismic behavior of SPSW structures were considered in the tests. The ultimate bearing capacity of SPSW structures with holes was estimated by AIJ codes and Ono reduction rate calculation methods. The major conclusions are indicated as follows:

(1)   SPSW structures without holes mainly develop failures at the roots of shear walls, accompanied with the concrete and bending failures of steel plates. SPSW structures with small holes also develop failures at the roots, but several steel plates crack along the corners of the holes. The failure mode is brittle to a certain extent. SPSW structures with large holes develop significant deformation and better ductility than SPSW structures with small holes but low ultimate loads.

(2)   Concrete cracks early under high axial compression ratio. The ultimate strength of SPSW structures increases, but the deformation capacity and ductility decrease to a certain extent. Deformation capacity is enhanced, but the ultimate bearing capacity decreases as hole size increases. With an increase in the thickness of stiffening plates, the ultimate bearing capacity of SPSW specimens increases. Eccentric holes decrease the earthquake-resistant energy-dissipation capacity of SPSW structures, and such a reduction is disadvantageous to seismic design. Therefore, eccentric holes on SPSW structures should be avoided.

(3)   The calculated results of AIJ codes are safe, but those of the Ono formula have high accuracy because it considers the influences of loading directions. In summary, both calculation formulas are feasible to calculate the ultimate shear capacity of SPSW structures with holes.

Author Contributions: Z.C. and J.W. carried out the experimental research and the academic paper writing. H.L. provided assistance during the experimental research.

Funding: This research received no external funding.

Conflicts of Interest: There is no conflict of interest.

References

1.   Akiyama, H.; Sekimoto, H.; Tanaka, M.; Inoue, K.; Fukihara, M.; Okuda, Y. 1/10th scale model test of inner concrete structure composed of concrete filled steel bearing wall. In Proceedings of the 10th SMi RT Conference H73, SMiRT 10, Anaheim, CA, USA, 22–27 August 1989; pp. 73–78.

2.   Lubell, A.S.; Prion, H.G.; Ventura, C.E.; Rezai, M. Unstiffened steel plate shear wall performance under cyclic loading. *J. Struct. Eng.* **2000**, *124*, 453–460. [CrossRef]

3.   USNRC Regulatory Guide 1.142 Revision 2. In *Safety-Related Concrete Structures for Nuclear Power Plants (Other than Reactor Vessels and Containments)*; NRC: Washington, DC, USA, 2001.

4.   Japanese Association of Architectural Technicians. *100 Typical Examples of Japanese Structural Technology*; Architecture and Building Press: Beijing, China, 2005.

5.   Berman, J.W.; Celik, O.C.; Bruneau, M. Comparing hysteretic behavior of light-gauge steel plate shear walls and braced frames. *Eng. Struct.* **2005**, *27*, 475–485. [CrossRef]

6.   Gan, C.; Lu, X.; Wang, W. Experimental study on the steel plate reinforced concrete shear walls. In Proceedings of the International Conference on Advances in Experimental Structural Engineering, Tongji University, Shanghai, China, 4–6 December 2007; pp. 176–188.

7.   Zhao, W.Y.; Guo, Q.Q.; Huang, Z.Y.; Tan, L.; Chen, J.; Ye, Y. Hysteretic model for steel–concrete composite shear walls subjected to in-plane cyclic loading. *Eng. Struct.* **2016**, *106*, 461–470. [CrossRef]

8. Wang, W.; Wang, Y.; Lu, Z. Experimental study on seismic behavior of steel plate reinforced concrete composite shear wall. *Eng. Struct.* **2018**, *160*, 281–292. [CrossRef]

9. Li, X.H.; Li, X.J. Steel plates and concrete filled composite shear walls related nuclear structural engineering: Experimental study for out-of-plane cyclic loading. *Nucl. Eng. Des.* **2017**, *315*, 144–154. [CrossRef]

10. Huang, S.T.; Huang, Y.S.; He, A.; Tang, X.; Chen, Q.; Liu, X.; Cai, J. Experimental study on seismic behavior of an innovative composite shear wall. *J. Construct. Steel Res.* **2018**, *148*, 165–179. [CrossRef]

11. Rafiei, S.; Hossain, K.M.A.; Lachemi, M.; Behdinan, K.; Anwar, M.S. Finite element modeling of double skin profiled composite shear wall system under in-plane loadings. *Eng. Struct.* **2013**, *56*, 46–57. [CrossRef]

12. Hu, H.S.; Nie, J.G.; Eatherton, M.R. Deformation capacity of concrete-filled steel plate composite shear walls. *J. Constr. Steel Res.* **2014**, *103*, 148–158. [CrossRef]

13. Booth, P.N.; Varma, A.H.; Sener, K.C.; Mori, K. Seismic behavior and design of a primary shield structure consisting of steel-plate composite (SC) walls. *Nucl. Eng. Des.* **2015**, *295*, 829–843. [CrossRef]

14. Nguyen, N.H.; Whittaker, A.S. Numerical modelling of steel-plate concrete composite shear walls. *Eng. Struct.* **2017**, *150*, 1–11. [CrossRef]

15. Wang, B.; Jiang, H.; Lu, X. Seismic performance of steel plate reinforced concrete shear wall and its application in China Mainland. *J. Construct. Steel Res.* **2017**, *131*, 132–143. [CrossRef]

16. Sato, K. Kitano Masato Experimental Study on Steel Plate Reinforced Concrete Structure Part 23 Bending Shear Tests (Test Results for SC Wall with Opening). In *Summary of Academic Speeches*; Japanese Architectural Society: Toyko, Japan, 1999; pp. 1227–1228.

17. Satou, K. Experimental Study on Steel Plate Reinforced Concrete Structure Part 24 Bending Shear Tests (Analysis of SC wall with Opening). In *Summary of Academic Speeches*; Japanese Architectural Society: Toyko, Japan, 1999; pp. 1229–1230.

18. Ishida, M. Experimental Study on Steel Plate Reinforced Concrete Structure Part 25 Bending Shear Tests (Evaluation Method for SC wall with Opening). In *Summary of Academic Speeches*; Japanese Architectural Society: Toyko, Japan, 1999; pp. 1231–1232.

19. Fujita, T. Experimental Study on Steel Plate Reinforced Concrete Structure Part 26 Shear Tests (Outline of the Experimental Study on SC Panels having Opening). In *Summary of Academic Speeches*; Japanese Architectural Society: Toyko, Japan, 1999; pp. 123–1234.

20. Osuka. Experimental Study on Steel Plate Reinforced Concrete Structure Part 27 Shear Tests (Discussion of Analytical Results and Experimental ones on SC Panels having Opening). In *Summary of Academic Speeches*; Japanese Architectural Society: Toyko, Japan, 1999; pp. 1235–1236.

21. Khatir, S.; Wahab, M.A. A computational approach for crack identification in plate structures using XFEM, XIGA, PSO and Jaya algorithm. *Theor. Appl. Fract. Mech.* **2019**, *103*, 102240. [CrossRef]

22. Liang, Y.C.; Sun, Y.P. Hole/Crack Identification in Circular Piezoelectric Plates. *Procedia Eng.* **2014**, *79*, 194–203. [CrossRef]

23. Khatira, S.; Wahabb, M.A. Fast simulations for solving fracture mechanics inverse problems using POD-RBF XIGA and Jaya algorithm. *Eng. Fract. Mech.* **2019**, *205*, 285–300. [CrossRef]

24. *Japanese Code (JEAC 4618-2009). Technical Specification for Seismic Design of Steel Plate Concrete Structures*; Japan Electric Association: Toyko, Japan, 1999.

25. Industrial Standard of the People's Republic of China. *Specification for Seismic Test of Building: JGJ/T 101-2015*; Ministry of Housing and Urban-Rural Development of the People's Republic of China: Beijing, China, 2015.

26. Japanese Architectural Society (AIJ). *The Calculation Standards and Specification for Reinforced Concrete Structure*; Japan Electric Association: Toyko, Japan, 1999.

27. Ono, M. NISIYAMA taku. Experimental Study on Reinforced Concrete Opening Wall above Opening Periphery Ratio 0.4: Part 2. Area Expansion, Ultimate Strength. In *Summary of Academic Speeches*; Japanese Architectural Society: Japan, Tokyo, 1995; pp. 147–150.

Article

Calibration Factor for ASCE 41-17 Modeling Parameters for Stocky Rectangular RC Columns

Sang Whan Han [1], Chang Seok Lee [1,*], Mary Ann Paz Zambrana [1] and Kihak Lee [2]

[1]   Department of Architectural Engineering, Hanyang University, Seoul 04763, Korea;
      swhan82@gmail.com (S.W.H.); arq.ann.paz@gmail.com (M.A.P.Z.)
[2]   Department of Architectural Engineering, Sejong University, Seoul 05006, Korea; kihaklee@sejong.ac.kr
*   Correspondence: mtsonicc@gmail.com; Tel.: +82-2-2220-0566

Received: 15 October 2019; Accepted: 26 November 2019; Published: 29 November 2019

Featured Application: Calibration factor for ASCE 41-17 modeling parameters of stocky columns.

Abstract: Existing old reinforced concrete (RC) buildings could be vulnerable to large earthquake events. Most columns in such buildings have insufficient reinforcement details, which may experience failure during an early loading stage. The failure of columns may lead to partial or complete collapse of entire building systems. To prepare for an adequate retrofit plan for columns, it is necessary to simulate the cyclic behavior of columns using a numerical model with adequate values of constituent modeling parameters. The nonlinear component modeling parameters are specified in ASCE 41-17. However, the experiments on stocky RC columns suggest that ASCE 41-17 nonlinear component modeling parameters do not reflect the RC column behavior adequately. To accurately simulate the nonlinear load–deformation responses of stocky RC columns with low span-to-depth ratio, this study proposes a calibration factor for ASCE 41-17 RC column modeling parameters. For this purpose, this study collected test data of 47 stocky column specimens. Based on the test data, empirical equations including the calibration factor for modeling parameters "*a*" and "*b*" in ASCE 41-17 were proposed. The accuracy of the proposed equation was verified by comparing the measured and calculated envelope curves.

Keywords: columns; cyclic behavior; low height-to-depth ratio; modeling parameters; calibration

---

1. Introduction

Reinforced concrete (RC) buildings not designed according to modern seismic design codes can be vulnerable to collapse during earthquakes [1]. Postearthquake researches indicated that columns are one of the most critical structural components in seismically active regions due to their nonductile reinforcement details [2–6]. Deficiencies such as lack of confinement due to widely spaced transverse reinforcement [1], ineffective anchorage length, and insufficient development length [7–11] can cause premature shear failure in RC columns [12,13], thus leading to rapid lateral strength degradation [14] or even collapse [15–18]. To understand the seismic behavior of such columns under earthquakes and to propose an adequate retrofit scheme, the seismic capacity of RC columns with nonseismic reinforcement details should be estimated.

To predict the seismic capacity of such RC columns with acceptable accuracy, an accurate backbone curve is required, which defines the lateral load–deformation capacity of a structural component. An extensive amount of research has been carried out on developing backbone curves of RC columns [14,15,19–22]. The nonlinear load–deformation relation proposed by ASCE 41-17 [23] (as noted hereafter as ASCE 41-17) is most widely used in the fields. This standard is provided to conduct seismic evaluation and retrofit of existing building by the American Society of Civil Engineer

(ASCE), which includes performance objectives and seismic hazards, Tiers 1–3 seismic performance evaluation procedures, and retrofit.

Due to the importance of this standard, it has been continuously revised since its first release to provide a better accuracy in estimating nonlinear load–deformation response of structural components.

Ghannoum and Matamoros [24] proposed equations for RC columns to calculate modeling parameters to construct a backbone curve, instead of using fixed values used in the previous standard [25]. These equations are shown in ASCE 41-17, which lead to better estimation of modeling parameters compared with those in the previous standard. Accurate estimation of modeling parameters is a key factor in seismic performance evaluation [26] by selecting input seismic ground motions [27].

However, the deformation parameters suggested in ASCE 41-17 do not consider height-to-depth ratios, which can significantly alter the column behavior [19,28]. For RC members with a height-to-depth ratio lower than 2 (as noted hereafter as stocky columns), their cyclic behavior is always dominated by shear [5,29,30]. Therefore, shear deformation induced by shear cracking expansion can be significant [31,32]. When the ASCE 41-17 modeling parameter equations are applied to columns with low height-to-depth ratios (stocky columns), the estimation error between measured and estimated values may become significant.

In order to better estimate the load–deformation response of stocky RC columns, this study proposes a calibration factor for the equations of deformation modeling parameters provided in ASCE 41-17. For this purpose, the values of deformation parameters for 47 rectangular RC columns with low height-to-depth ratios are extracted. To minimize the difference between measured values and those predicted by the ASCE 41-17 equations, the calibration factor was estimated for individual column specimens. Then, the empirical equations including the calibration factor are proposed from regression analyses.

2. Backbone Curve for RC Columns in ASCE 41-17

In ASCE 41-17, an idealized nonlinear load–deformation relation for a structural component is provided to represent a generalized backbone curve, as shown in Figure 1. In Figure 1, V and V_y are the column shear force and the column yield shear strength, respectively, and θ is the drift ratio; a and b are the parameters representing plastic drift ratios, for cases when a strength degradation begins and the shear force drops to the point "D", respectively.

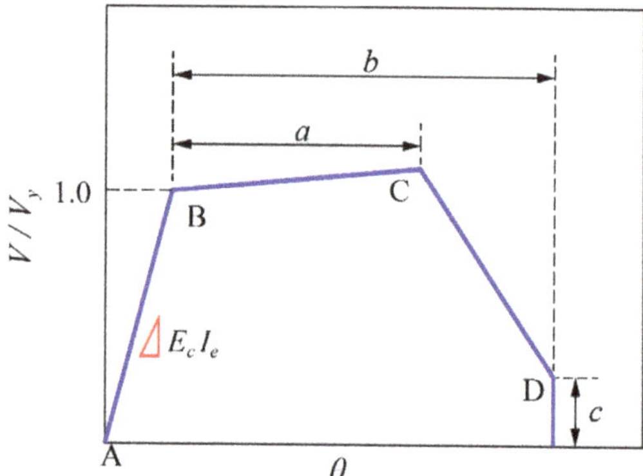

Figure 1. Generalized load–deformation relation.

The generalized load–deformation relation consists of a linear portion and a nonlinear portion. The linear portion corresponds to the segment between points A and B, which can be determined using the effective stiffness of a structural component. In ASCE 41-17, equations are given to calculate the effective stiffness for various structural components. For RC columns, the proposed equations to calculate the effective stiffness are summarized in Table 1, where E_c is the modulus of elasticity of concrete, I_g is the moment of inertia of the concrete gross section, A_w is the summation of the net horizontal cross-sectional area for concrete in the direction of loading, and A_g is the gross sectional area of the column.

Table 1. Effective stiffness equations proposed by ASCE 41-17.

Component	Flexural Rigidity	Shear Rigidity	Axial Rigidity
Columns with compression caused by design gravity loads	$0.7E_{cE}I_g$	$0.4E_{cE}A_w$	$E_{cE}A_g$
Columns with compression caused by design gravity loads or with tension	$0.3E_{cE}I_g$	$0.4E_{cE}A_w$	$0.4E_{cE}A_g$

The nonlinear portion is governed by two deformation parameters, that is, modeling parameters a and b. For rectangular RC columns, parameter a can be calculated using Equation (1):

$$a_{ASCE} = 0.042 - 0.043\frac{P}{A_g f_c} + 0.63\rho_t - 0.023\frac{V_y}{V_0} \geq 0, \tag{1}$$

where P is the axial compressive load applied on the column section, f_c is the concrete compressive strength, ρ_t is the ratio of the area of distributed transverse reinforcement to the concrete gross area perpendicular to that reinforcement, and V_y is the yield shear strength of the column. The column shear strength V_0 is calculated from Equation (2) provided in ASCE 41-17 for RC columns:

$$V_0 = \left[\frac{A_v f_{yt} d}{s} + \left(\frac{6\sqrt{f_c}}{a_v/h}\sqrt{1 + \frac{P}{6A_g\sqrt{f_c}}}\right)0.8A_g\right], \tag{2}$$

where a_v/h is the height-to-depth ratio of the column; k_{nl} is 1.0 for a displacement ductility demand (μ) less than or equal to 2 and 0.7 for $\mu \geq 6$, where μ is calculated as θ/θ_y and θ_y is the yield drift ratio as shown in Figure 2. For μ between 2 and 6, k_{nl} is estimated using interpolation.

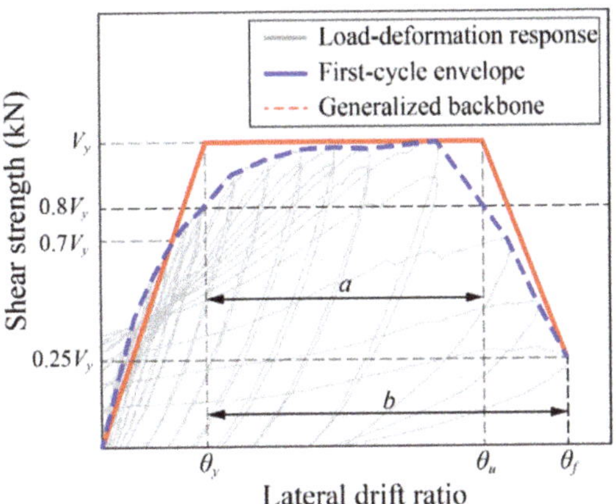

Figure 2. Measured modeling parameters from a measured first-cycle envelope.

The load–deformation response representing a reduced resistance is predicted using modeling parameter b, which is calculated using Equation (3):

$$b_{ASCE} = \frac{0.5}{5 + \frac{P}{0.8A_g f_c} \frac{1}{\rho_t} \frac{f_c}{f_{yt}}} - 0.01 \geq a_{ASCE},$$

(3)

where f_{yt} is the yield strength of the transverse reinforcement and modeling parameter b must be greater than or equal to modeling parameter a.

3. Estimating the Values of Modeling Parameters from the Measured Cyclic Curves

The modeling parameters a and b were extracted from the first-cyclic envelope curve of RC columns. The first-cycle envelope curve can be constructed by connecting each point of the peak displacements during the first cycle of each increment of loading (or deformation) [23,33]. Figure 2. illustrates the extraction of the modeling parameters a and b.

In Figure 2, the values for the maximum shear strength V_y are identical to the maximum ordinate values of the first-cycle envelope. Yield drift ratio θ_y was obtained according to the procedure proposed by Sezen et al. [2]. A secant line is projected from the origin to the intersection point on the first-cycle envelope curve and a horizontal line at $0.7V_y$. The secant line was extended, until it reached the horizontal line drawn at V_y. Then, θ_y is the abscissa of the intersection of these two lines.

Drift ratio θ_u is a lateral drift ratio, at which a significant (more than or equal to 20%) lateral resistance degradation from V_y occurs. In ASCE 41-17, θ_f is defined as a lateral drift ratio at axial failure. However, due to the scarcity of column specimens tested up to the onset of axial failure [24,34], θ_f is defined as a lateral drift ratio, when the lateral strength decreases to 25% of V_y.

The values of modeling parameters a and b can be calculated using Equations (4) and (5), respectively:

$$a = \theta_u - \theta_y,$$

(4)

$$b = \theta_f - \theta_y.$$

(5)

4. Stocky RC Column Database

The PEER Structural Performance Database provides the test data of 246 rectangular RC columns. Among them, only 47 columns have section-to-depth ratios less than or equal to 2.0. Only a limited number of cyclic tests were conducted for stock columns. In this study, 47 rectangular RC column test specimens with height-to-depth ratios (a_v/h) less than or equal to 2.0 (stocky columns) were collected from the PEER Structural Performance Database [35]. All the 47 specimens experienced a strength drop by more than 20%.

The ranges of important parameters are summarized below:

Height-to-depth ratio	$1.0 \leq a_v/h \leq 2.0$
Section width (mm)	$160 \leq b \leq 500$
Section height (mm)	$160 \leq h \leq 914$
Center-to-center spacing of a transverse reinforcement (mm)	$20 \leq s \leq 406$
Longitudinal reinforcement ratio (%)	$1.27 \leq \rho_l \leq 6.94$
Transverse reinforcement ratio (%)	$0.08 \leq \rho_t \leq 1.64$
Measured compressive strength of concrete	$14 \leq f_c \leq 118$
Measured yield strength of longitudinal reinforcement	$323 \leq f_{yl} \leq 510$
Measured yield strength of transverse reinforcement	$258 \leq f_{yt} \leq 1424$
Axial load ratio (%)	$0 \leq v \leq 80.1; v = P/\left(A_g f_c\right)$

To estimate the error between the measured and calculated values of the modeling parameters, absolute relative errors (*AREs*) were calculated using Equation (6):

$$ARE = \left| \frac{MP_{meas} - MP_{ASCE}}{MP_{meas}} \right|, \tag{6}$$

where MP_{meas} is the measured modeling parameter value for a and b extracted from the first-cycle envelope curve and MP_{ASCE} is the parameter value for a and b calculated using Equations (1) and (3), respectively.

Figure 3 shows the relationship of the calculated *ARE* and the shear span-to-section height ratio (a_v/h), in which 246 specimens with a_v/h ranging from 1 to 7.3 were included. The test data for these specimens were also obtained from the PEER Structural Performance Database [35]. The solid and dashed lines in Figure 3 represent the mean *ARE* values obtained from the moving windows analyses. With an increase in a_v/h, the mean values of *ARE* for parameters a and b [$\mu_{ARE(a)}, \mu_{ARE(b)}$] generally decrease. In the case of $a_v/h \geq 2.0$, $\mu_{ARE(a)}$ and $\mu_{ARE(b)}$ approach approximately 0.5 and 0.3, respectively. However, with the decrease in a_v/h within the range of a_v/h less than 2.0, $\mu_{ARE(a)}$ and $\mu_{ARE(b)}$ increase sharply, which indicates that the errors associated with the ASCE 41-17 equations for parameters a and b become more significant. Thus, it is necessary to propose a calibration factor for Equations (1) and (3) for columns with $a_v/h \leq 2.0$.

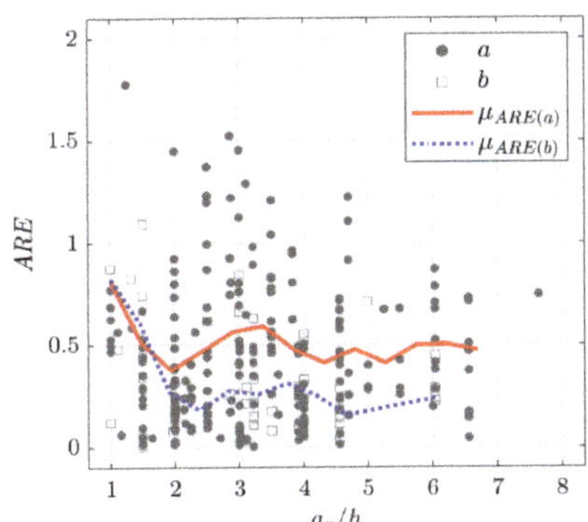

Figure 3. Absolute relative error (*ARE*) between the measured and ASCE modeling parameters.

5. Regression Analysis

The original equations of ASCE 41-17 do not include the influence of a height-to-depth ratio; however, it was revealed in the previous section that there was a considerable amount of error in the modeling parameter predictive equation proposed by ASCE 41-17 when the height-to-depth ratio was lower than 2.0. For this reason, a calibration factor was proposed to minimize error.

Linear regression was used to introduce a calibration factor for the modeling parameter predictive equation of ASCE 41-17 to calculate more accurately the values of parameters a and b. The measured modeling parameters extracted from the first-cycle envelope curve is shown in Table 2.

To propose an empirical calibration factor, various candidate predictor variables frequently used in the past research [14,22,52–55] were considered. Considered candidate predictor variables include: longitudinal reinforcement ratio (ρ_l), transverse reinforcement ratio (ρ_t), axial load ratio ($v = P/(f'_c A_g)$), measured concrete compressive strength (f_c), column shear span (a_v), height-to-depth ratio (a_v/h), ratio of a transverse bar spacing to a column depth (s/h), ratio of a strength contribution of longitudinal reinforcements to that of concrete ($f_{yl}A_{sl}/f'_c A_g$), ratio of a strength contribution of concrete axial strength to that of transverse reinforcements ($f'_c A_g/f_{yt}A_{st}$).

Table 2. Physical properties and measured deformation parameters of the selected specimens.

No.	ID	a_v/h	b (mm)	h (mm)	s (mm)	ρ_l (%)	ρ_t (%)	f_c (MPa)	f_{yt} (MPa)	a_{meas} (%)	b_{meas} (%)	Reference
1	SC9	1.33	457	914	406	1.88	0.08	16	400	0.36	-	[36]
2	CB060C	1.16	278	278	52	2.75	0.78	46	414	0.50	-	[37]
3	H-2-1_3	2.00	200	200	40	2.65	0.71	23	364	1.51	-	
4	H-2-1_5	2.00	200	200	50	2.65	0.57	23	364	1.96	-	[38]
5	HT-2-1_3	2.00	200	200	60	2.65	0.71	20	364	2.03	-	
6	HT-2-1_5	2.00	200	200	75	2.65	0.57	20	364	1.99	-	
7	I18	2.00	500	500	254	2.12	0.20	33	258	1.64	-	[39]
8	I21	2.00	500	500	254	2.12	0.20	32	258	0.98	-	
9	HPRC10-63	1.50	200	200	35	1.27	0.68	22	344	1.33	-	[40]
10	HPRC19-32	1.50	200	200	20	1.27	1.19	21	344	1.15	-	
11	N-18M	1.50	300	300	100	2.68	0.21	27	380	0.62	10.16	
12	N-27C	1.50	300	300	100	2.68	0.21	27	380	0.57	3.00	[41]
13	N-27M	1.50	300	300	100	2.68	0.21	27	380	0.94	4.70	
14	2D16RS	2.00	200	200	50	2.01	0.48	32	316	2.85	-	[42]
15	4D13RS	2.00	200	200	50	2.65	0.48	30	316	1.00	-	
16	CA025C	1.50	200	200	70	2.13	0.81	26	426	2.10	-	[43]
17	CA060C	1.50	200	200	70	2.13	0.81	26	426	0.81	-	
18	C1	1.50	300	300	160	1.69	0.08	14	587	0.46	2.33	
19	C12	1.50	300	300	75	1.69	0.28	18	384	1.21	8.15	[44]
20	C4	1.50	300	300	75	1.69	0.28	14	587	1.08	5.57	
21	C8	1.50	300	300	75	1.69	0.28	18	384	0.74	2.21	
22	D1	1.00	300	300	50	1.69	0.42	28	398	0.73	3.96	
23	D11	1.50	300	300	150	2.25	0.14	28	398	0.66	1.89	
24	D12	1.50	300	300	150	2.25	0.14	28	398	0.69	1.98	[45]
25	D14	1.50	300	300	50	2.25	0.42	26	398	1.49	17.98	
26	D16	1.00	300	300	50	1.69	0.42	26	398	0.75	8.45	
27	B3	2.00	250	250	60	2.43	0.63	100	344	1.11	-	
28	B4	2.00	250	250	60	2.43	0.52	100	1126	2.34	-	
29	B5	2.00	250	250	30	2.43	1.05	100	774	1.18	-	[46]
30	B6	2.00	250	250	60	2.43	0.52	100	857	1.32	-	
31	B7	2.00	250	250	30	1.81	0.52	100	774	0.54	-	
32	UC15H	2.00	225	225	45	1.86	1.27	118	1424	1.40	-	
33	UC15L	2.00	225	225	45	1.86	1.27	118	1424	2.36	-	[47]
34	UC20H	2.00	225	225	35	1.86	1.63	118	1424	2.30	-	
35	UC20L	2.00	225	225	35	1.86	1.63	118	1424	2.93	-	
36	CUS	1.11	230	410	89	3.01	0.55	35	414	0.69	-	
37	CUW	1.98	410	230	89	3.01	0.15	35	414	1.21	-	[48]
38	UM207	2.00	200	200	100	1.99	0.28	18	324	0.81	-	
39	HC4-8L16-T10-0.2P	2.00	254	254	51	2.46	1.64	86	510	5.84	-	[49]
40	No.1	2.00	300	300	100	2.68	0.19	31	392	0.64	-	
41	No.3	2.00	300	300	200	2.68	0.09	31	392	0.55	8.00	[50]
42	No.4	2.00	300	300	100	2.68	0.19	31	392	0.62	1.95	
43	BE	1.00	175	175	110	2.42	0.29	33	312	0.28	2.01	
44	CE	1.00	175	175	110	2.42	0.29	26	312	0.28	-	[51]
45	LE	1.00	175	175	20	6.94	1.62	42	322	0.77	-	
46	104-08	1.00	160	160	40	2.22	0.61	20	559	0.56	-	
47	204-08	2.00	160	160	40	2.22	0.61	21	559	1.10	-	[29]

Among the candidate variables listed above, only the statistically significant variables used to predict the deformation parameters were selected. To determine the statistically significant variables, a linear stepwise regression analysis [56] was used. A candidate variable is considered to be statistically significant, when the p-value is less than 5%. The p-value is used for measuring the plausibility of a null hypothesis. Typically, in regression analysis, a null hypothesis is rejected if the p-value is less than 0.05.

As a result of the regression analysis, the obtained equation for parameter a was written as:

$$a_{fit} = a_{ASCE} + \left(-\frac{1}{47.78} - \frac{1}{57.49}\frac{1}{a_v/h} + \frac{f_{yl}}{1988.07 c_{unit}} \right), \tag{7}$$

where c_{unit} is 6.89 when using the unit of MPa and 1.00 when using the unit of ksi.

As discussed in the previous section, height-to-depth ratio a_v/h was found to be a significant predictor variable.

The equation obtained for parameter b was described as:

$$b_{fit} = b_{ASCE} + \left(\frac{1}{12.56} - \frac{1}{703.39 \rho_l} \right) \geq a_{fit}. \tag{8}$$

In the proposed equations (Equations (7) and (8)), three variables were considered (a_v/h, ρ_l, and f_{yl}). It was reported that RC member behavior is affected significantly by a_v/h [1–6,57]. RC columns with lower a_v/h are more likely to fail in shear. Past studies [7–9] also reported that the dowel action of longitudinal reinforcements significantly affects the load–deformation responses of stocky structural components; thus, ρ_l and f_{yl} selected from the stepwise regression analyses can be physically meaningful variables.

The significant predictor variable for both modeling parameters a and b was related to the longitudinal reinforcement, that is, the measured yield strength of a longitudinal reinforcement f_{yl} and the longitudinal reinforcement ratio ρ_l. Past literatures [58–60] also revealed that dowel action of longitudinal reinforcements can have significant impact on load–deformation response of relatively stocky structural components.

As discussed in the previous section, due to the scarcity of column specimens tested to the onset of axial failure, Equation (8) was proposed for column specimens experiencing a 75% lateral resistance degradation in Table 2, which corresponded to the 15 RC column specimens.

6. Validation of the Proposed Calibration Factor

Figure 4 shows the accuracy of the calculated values of modeling parameters a and b with respect to the corresponding measured values. As shown in Figure 4, the calculated values of parameters a and b by applying the calibration factor proposed by this study match the corresponding measured values more accurately than those obtained without applying the calibration factor. The errors associated with parameters a and b were calculated using Equation (6).

The values of *Error* associated with the calculated values of parameter a without applying the proposed calibration factor is 87.7%, whereas that by applying the calibration factor is 41.4%. Similar results were obtained for parameter b.

Envelope curves were plotted for specimens, which were obtained using the ASCE 41-17 equations for parameters a and b with and without the proposed calibrating factor applied. Figure 5 show the envelope curves and the measured cyclic curves for the specimens No.4, C1, C8, and N27-C. It was observed that the envelope curves obtained with the proposed calibration factor match the measured cyclic curves more accurately than those obtained without the calibration factor. Similar observations were obtained for the other specimens listed in Table 2, which were not included in this paper due to page limitations.

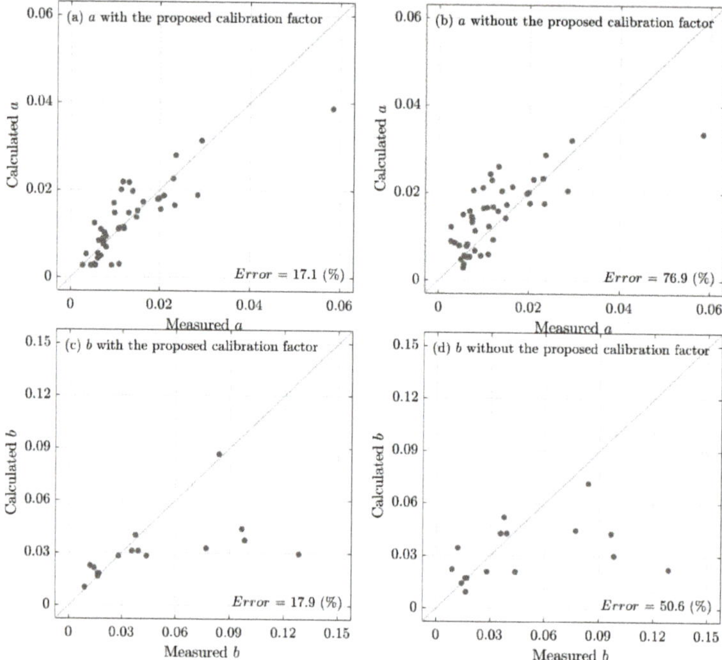

Figure 4. Accuracies of the calculated values of modeling parameter a by applying the proposed calibration factor (**a**) and without applying the proposed calibration factor (**b**). Accuracies of the calculated values of modeling parameter b by applying the proposed calibration factor (**c**) and without applying the proposed calibration factor (**d**).

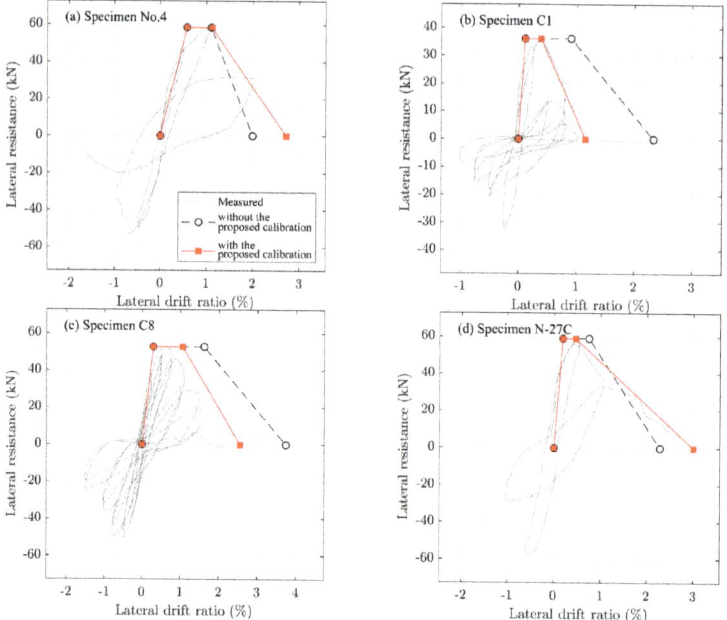

Figure 5. Predicted envelope curves and measured cyclic curves.

7. Conclusions

In this study, a database of 47 stocky RC columns was used to extract the modeling parameters for constructing column envelope curves. The extracted values were then compared with the modeling parameters values calculated using the ASCE 41-17 equations. The comparison results showed that the error between the measured values and the values calculated with the ASCE 41-17 equations was significant for the columns with height-to-depth ratios (a_v/h) less than 2.0. In this study, a calibration factor was proposed for RC columns with a_v/h less than 2.0 to improve the accuracy of the values of modeling parameters a and b calculated from the ASCE 41-17 equations. The following conclusion can be drawn.

1. The ASCE 41-17 equations for deformation modeling parameters produce erroneous predictions for stocky RC columns, because these equations do not consider the effect of a_v/h of RC columns. The application of the proposed calibration factor significantly improved the accuracy of the calculated values of modeling parameters a and b for columns with a_v/h less than 2.0.
2. After applying the calibration factor to the ASCE 41-17 equations, errors in the calculated values of modeling parameters a and b significantly decreased. For parameter a, the error reduced from 87.7% to 41.4%, and for parameter b, the error reduced from 71.1% to 42.3%.
3. The envelope curves for the stocky RC columns were accurately constructed using the modeling parameters with the proposed calibration factor, which match the measured cyclic curves.

Author Contributions: Conceptualization, S.W.H.; methodology, C.S.L.; investigation, C.S.L., M.A.P.Z., K.L.; writing of the original draft preparation, S.W.H., M.A.P.Z.; writing of review and editing, S.W.H.; visualization, C.S.L.; supervision, S.W.H.; project administration, S.W.H.

Funding: This research was funded by the Korea Agency for Infrastructure Technology Advancement (grant number 19CTAP-C152179-01).

Acknowledgments: The research was supported by a grant from the Korea Agency for Infrastructure Technology (19CTAP-C152179-01). The valuable comments of four anonymous reviewers are greatly appreciated.

Conflicts of Interest: The authors declare no conflicts of interest.

References

1. Lynn, A.C.; Moehle, J.P.; Mahin, S.A.; Holmes, W.T. Seismic evaluation of existing reinforced concrete building columns. *Earthq. Spectra* **1996**, *12*, 715–739. [CrossRef]
2. Sezen, H.; Whittaker, A.S.; Elwood, K.J.; Mosalam, K.M. Performance of reinforced concrete buildings during the August 17, 1999 Kocaeli, Turkey earthquake, and seismic design and construction practise in Turkey. *Eng. Struct.* **2003**, *25*, 103–114. [CrossRef]
3. Humar, J.M.; Lau, D.; Pierre, J.-R. Performance of buildings during the 2001 Bhuj earthquake. *Can. J. Civ. Eng.* **2001**, *28*, 979–991. [CrossRef]
4. Kam, W.Y.; Pampanin, S.; Dhakal, R.; Gavin, H.P.; Roeder, C. Seismic performance of reinforced concrete buildings in the September 2010 Darfield (Canterbury) earthquake. *Bull. N. Z. Soc. Earthq. Eng.* **2010**, *43*, 340–350. [CrossRef]
5. Galal, K.; Arafa, A.; Ghobarah, A. Retrofit of RC square short columns. *Eng. Struct.* **2005**, *27*, 801–813. [CrossRef]
6. Liel, A.B.; Haselton, C.B.; Deierlein, G.G. Seismic collapse safety of reinforced concrete buildings. II: Comparative assessment of nonductile and ductile moment frames. *J. Struct. Eng.* **2011**, *137*, 492–502. [CrossRef]
7. Lee, C.S.; Han, S.W. Cyclic behaviour of lightly-reinforced concrete columns with short lap splices subjected to unidirectional and bidirectional loadings. *Eng. Struct.* **2019**, *189*, 373–384. [CrossRef]
8. Harries, K.A.; Ricles, J.M.; Pessiki, S.; Sause, R. Seismic retrofit of lap splices in nonductile columns using CFRP jackets. *ACI Struct. J.* **2006**, *103*, 874–884. [CrossRef]

9. Melek, M.; Wallace, J. Performance of columns with short lap splices. In Proceedings of the 13th World Conference on Earthquake Engineering, Vancouver, BC, Canada, 1–6 August 2004.

10. Cho, J.-Y.; Pincheira, J.A. Nonlinear modeling of RC columns with short lap splices. In Proceedings of the 13th World Conference on Earthquake Engineering, Vancouver, BC, Canada, 1–6 August 2004.

11. Kim, C.-G.; Park, H.-G.; Eom, T.-S. Seismic performance of reinforced concrete columns with lap splices in plastic hinge region. *ACI Struct. J.* **2018**, *115*, 235–245. [CrossRef]

12. Sezen, H.; Moehle, J.P. Shear strength model for lightly reinforced concrete columns. *J. Struct. Eng.* **2004**, *130*, 1692–1703. [CrossRef]

13. Sezen, H.; Moehle, J.R. Seismic tests of concrete columns with light transverse reinforcement. *ACI Struct. J.* **2006**, *103*, 842–849. [CrossRef]

14. Elwood, K.J.; Moehle, J.P. Drift capacity of reinforced concrete columns with light transverse reinforcement. *Earthq. Spectra* **2005**, *21*, 71–89. [CrossRef]

15. Elwood, K.J.; Moehle, J.P. Axial capacity model for shear-damaged columns. *ACI Struct. J.* **2005**, *102*, 578–587. [CrossRef]

16. Baradaran Shoraka, M.; Elwood, K.J. Mechanical model for non ductile reinforced concrete columns. *J. Earthq. Eng.* **2013**, *17*, 937–957. [CrossRef]

17. Elwood, K.J.; Moehle, J.P. Dynamic collapse analysis for a reinforced concrete frame sustaining shear and axial failures. *Earthq. Eng. Struct. Dyn.* **2008**, *37*, 991–1012. [CrossRef]

18. Moehle, J.P.; Mahin, S.A. Observations on the behavior of reinforced concrete buildings during earthquakes. *ACI Symp. Publ.* **1991**, *127*, 67–90. [CrossRef]

19. Haselton, C.B.; Liel, A.B.; Taylor-Lange, S.C.; Deierlein, G.G. Calibration of model to simulate response of reinforced concrete beam-columns to collapse. *ACI Struct. J.* **2016**, *113*, 1141–1152. [CrossRef]

20. Sezen, H.; Chowdhury, T. Hysteretic model for reinforced concrete columns including the effect of shear and axial load failure. *J. Struct. Eng.* **2009**, *135*, 139–146. [CrossRef]

21. Ghannoum, W.M.; Moehle, J.P. Rotation-based shear failure model for lightly confined RC columns. *J. Struct. Eng.* **2012**, *138*, 1267–1278. [CrossRef]

22. Elwood, K.J.; Eberhard, M.O. Effective stiffness of reinforced concrete columns. *ACI Struct. J.* **2009**, *106*, 476–484. [CrossRef]

23. ASCE 41-17. *Seismic Evaluation and Retrofit of Existing Buildings*; American Society of Civil Engineers: Reston, VA, USA, 2017. [CrossRef]

24. Ghannoum, W.M.; Matamoros, A.B. Nonlinear modeling parameters and acceptance criteria for concrete columns. *ACI Spec. Publ.* **2014**, *297*, 1–24.

25. ASCE 41-13. *Seismic Evaluation and Retrofit of Existing Buildings*; American Society of Civil Engineers: Reston, VA, USA, 2014. [CrossRef]

26. Wen, Y.K.; Collins, K.R.; Han, S.W.; Elwood, K.J. Dual-level designs of buildings under seismic loads. *Struct. Saf.* **1996**, *18*, 195–224. [CrossRef]

27. Lee, L.H.; Lee, H.H.; Han, S.W. Method of selecting design earthquake ground motions for tall buildings. *Struct. Des. Tall Build.* **2000**, *9*, 201–213. [CrossRef]

28. Elwood, K.J.; Matamoros, A.B.; Wallace, J.W.; Lehman, D.E.; Heintz, J.A.; Mitchell, A.D.; Moore, M.A.; Valley, M.T.; Lowes, L.N.; Comartin, C.D.; et al. Update to ASCE/SEI 41 concrete provisions. *Earthq. Spectra* **2007**, *23*, 493–523. [CrossRef]

29. Zhou, X.; Satoh, T.; Jiang, W.; Ono, A.; Shimizu, Y. Behavior of reinforced concrete short column under high axial load. *Trans. Jpn. Concr. Inst.* **1987**, *9*, 541–548.

30. Han, S.W.; Lee, C.S.; Shin, M.; Lee, K. Cyclic performance of precast coupling beams with bundled diagonal reinforcement. *Eng. Struct.* **2015**, *93*, 142–151. [CrossRef]

31. Chen, Z.; Ding, Y.; Shi, Y.; Li, Z. A vertical isolation device with variable stiffness for long-span spatial structures. *Soil Dyn. Earthq. Eng.* **2019**, *123*, 543–558. [CrossRef]

32. Moretti, M.; Tassios, T.P. Behaviour of short columns subjected to cyclic shear displacements: Experimental results. *Eng. Struct.* **2007**, *29*, 2018–2029. [CrossRef]

33. FEMA P440A. *Effects of Strength and Stiffness Degradation on Seismic Response*; 440A; Department of Homeland Security: Washington, DC, USA, 2009.

34. Berry, M.P.; Eberhard, M.O. *Performance Models for Flexural Damage in Reinforced Concrete Columns*; Pacific Earthquake Engineering Research Center: Berkeley, CA, USA, 2004.

35. Berry, M.; Parrish, M.; Eberhard, M. *PEER Structural Performance Database User's Manual*; University of California, Berkeley: Berkeley, CA, USA, 2004.

36. Aboutaha, R.S.; Engelhardt, M.D.; Jirsa, J.O.; Kreger, M.E. Rehabilitation of shear critical concrete columns by use of rectangular steel jackets. *ACI Struct. J.* **1999**, *96*, 68–78. [CrossRef]

37. Amitsu, S.; Shirai, N.; Adachi, H.; Ono, A. Deformation of reinforced concrete column with high or fluctuating axial force. *Trans. Jpn. Concr. Inst.* **1991**, *13*, 355–362.

38. Esaki, F. Reinforcing effect of steel plate hoops on ductility of R/C square columns. In Proceedings of the 11th World Conference on Earthquake Engineering, Acapulco, Mexico, 23–28 June 1996.

39. Iwasaki, T.; Kawasima, K.; Hagiwara, R.; Koyama, T.; Yoshida, T. *Experimental Investigation on Hysteretic Behavior of Reinforced Concrete Bridge Pier Columns*; Public Works Research Institute: Tsukuba, Japan, 1985.

40. Nagasaka, T. Effectiveness of steel fiber as web reinforcement in reinforced concrete columns. *Trans. Jpn. Concr. Inst.* **1982**, *4*, 493–500.

41. Nakamura, T.; Yoshimura, M. Gravity load collapse of reinforced concrete columns with brittle failure modes. *J. Asian Arch. Build. Eng.* **2002**, *1*, 21–27. [CrossRef]

42. Ohue, M.; Morimoto, H.; Fujii, S.; Morita, S. The behavior of RC short columns failing in splitting bond-shear under dynamic lateral loading. *Trans. Jpn. Concr. Inst.* **1985**, *7*, 293–300.

43. Ono, A.; Shirai, N.; Adachi, H.; Sakamaki, Y. Elasto-plastic behavior of reinforced concrete column with fluctuating axial force. *Trans. Jpn. Concr. Inst.* **1989**, *11*, 239–246.

44. Ousalem, H.; Kabeyasawa, T.; Tasai, A.; Ohsugi, Y. Experimental study on seismic behavior of reinforced concrete columns under constant and variable axial loadings. In Proceedings of the Japan Concrete Institute, Tsukuba, Japan, June 2002.

45. Ousalem, H.; Kabeyasawa, T.; Tasai, A. Effect of hysteretic reversals on lateral and axial capacities of reinforced concrete columns. In Proceedings of the Fifth U.S.-Japan Workshop on Performance-Based Earthquake Engineering Methodology for Reinforced Concrete Structures, Hakone, Japan, 10–11 September 2003.

46. Sakai, Y.; Hibi, J.; Otani, S.; Aoyama, H. Experimental study on flexural behavior of reinforced concrete columns using high-strength concrete. *Trans. Jpn. Concr. Inst.* **1990**, *12*, 323–330.

47. Sugano, S. Seismic behavior of reinforced concrete columns which used ultra-high-strength concrete. In Proceedings of the 11th World Conference on Earthquake Engineering, Acapulco, Mexico, 23–28 June 1996.

48. Umehara, H.; Jirsa, J.O. *Shear Strength and Deterioration of Short Reinforced Concrete Columns Under Cyclic Deformations*; PMFSEL Report No. 82-3; University of Texas at Austin: Austin, TX, USA, 1982.

49. Xiao, Y.; Martirossyan, A. Seismic performance of high-strength concrete columns. *J. Struct. Eng.* **1998**, *124*, 241–251. [CrossRef]

50. Takaine, Y.; Yoshimura, M.; Nakamura, T. Collapse drift of reinforced concrete columns. *J. Struct. Constr. Eng. (Trans. AIJ)* **2003**, *68*, 153–160. [CrossRef]

51. Yoshimura, K.; Kikcuri, K.; Kuroki, M. Seismic shear strengthening method for existing R/C short columns. *ACI Spec. Publ.* **1991**, *128*, 1065–1080. [CrossRef]

52. LeBorgne, M.R.; Ghannoum, W.M. Calibrated analytical element for lateral-strength degradation of reinforced concrete columns. *Eng. Struct.* **2014**, *81*, 35–48. [CrossRef]

53. Han, S.W.; Koh, H.; Lee, C.S. Accurate and efficient simulation of cyclic behavior of diagonally reinforced concrete coupling beams. *Earthq. Spectra* **2019**, *35*, 361–381. [CrossRef]

54. Lee, C.S.; Han, S.W. Computationally effective and accurate simulation of cyclic behaviour of old reinforced concrete columns. *Eng. Struct.* **2018**, *173*, 892–907. [CrossRef]

55. Elwood, K.J. Modelling failures in existing reinforced concrete columns. *Can. J. Civ. Eng.* **2004**, *31*, 846–859. [CrossRef]

56. Chatterjee, S.; Hadi, A.S. *Regression Analysis by Example*, 4th ed.; Wiley Series in Probability and Statistics; Wiley: Hoboken, NJ, USA, 2006. [CrossRef]

57. Shin, M.; Gwon, S.-W.; Lee, K.; Han, S.W.; Jo, Y.W. Effectiveness of high performance fiber-reinforced cement composites in slender coupling beams. *Constr. Build. Mater.* **2014**, *68*, 476–490. [CrossRef]

58. Barney, G.B.; Shiu, K.N.; Rabbat, B.G.; Fiorato, A.E.; Russell, H.G.; Corley, W.G. *Behavior of Coupling Beams Under Load Reversals (RD068.01B)*; Portland Cement Association: Skokie, IL, USA, 1980.

59. Breña, S.F.; Ihtiyar, O. Performance of conventionally reinforced coupling beams subjected to cyclic loading. *J. Struct. Eng.* **2011**, *137*, 665–676. [CrossRef]
60. Ding, R.; Tao, M.-X.; Nie, J.-G.; Mo, Y.L. Shear deformation and sliding-based fiber beam-column model for seismic analysis of reinforced concrete coupling beams. *J. Struct. Eng.* **2016**, *142*, 04016032. [CrossRef]

Article

Development of an ANN-Based Lumped Plasticity Model of RC Columns Using Historical Pseudo-Static Cyclic Test Data

Zhenliang Liu [1,2] **and Suchao Li** [1,2,3,*]

1   Key Laboratory of Structural Dynamic Behavior and Control of Ministry of Education,
    School of Civil Engineering, Harbin Institute of Technology, Harbin 150090, China; 15B933009@hit.edu.cn
2   Key Lab of Smart Prevention and Mitigation of Civil Engineering Disasters of Ministry of Industry and
    Information Technology, Harbin Institute of Technology, Harbin 150090, China
3   Department of Civil Engineering, Harbin Institute of Technology at Weihai, Weihai 264209, China
*   Correspondence: lisuchao@hit.edu.cn; Tel.: +86-1363-360-1698

Received: 19 September 2019; Accepted: 5 October 2019; Published: 11 October 2019

Featured Application: Predicting the critical properties of RC columns using the developed ANN model, providing parameters for their numerical simulation based upon a lumped plasticity model.

Abstract: This study explores the possibility of using an ANN-based model for the rapid numerical simulation and seismic performance prediction of reinforced concrete (RC) columns. The artificial neural network (ANN) method is implemented to model the relationship between the input features of RC columns and the critical parameters of the commonly used lumped plasticity (LP) model: The strength and the yielding, capping and ultimate deformation capacity. Cyclic test data of 1163 column specimens obtained from the PEER and NEEShub database and other sources are collected and divided into the training set, test set and validation set for the ANN model. The effectiveness of the proposed ANN model is validated by comparing it with existing explicit formulas and experimental results. Results indicated that the developed model can effectively predict the strength and deformation capacities of RC columns. Furthermore, the response of two RC frame structures under static force and strong ground motion were simulated by the ANN-based, bi-linear and tri-linear LP model method. The good agreement between the proposed model and test results validated that the ANN-based method can provide sufficiently accurate model parameters for modeling the seismic response of RC columns using the LP model.

Keywords: strength; yielding capping and ultimate deformation; RC column; cyclic test database; artificial neural network; bi-linear and tri-linear lumped plasticity model

---

1. Introduction

Reinforced concrete (RC) columns are fundamental structural components that are widely used in civil infrastructures such as buildings and bridges. In seismic-prone regions, the seismic performance of such structural components significantly affects structural safety in seismic events. However, due to the nonlinearity of structural materials and the uncertainty of earthquake excitations, there are still some difficulties exists for researchers and engineers in some fields, e.g., the rapid seismic evaluation of a regional transport network, including a number of bridges [1]. Therefore, it is still necessary to develop a rapid and reliable model for the main components of structures, e.g., RC columns and piers.

Researchers and engineers in seismic engineering have conducted many laboratory experiments, including pseudo-static and shaking table tests, to investigate the mechanical properties of RC columns. Based on them, several explicit formulas have also been established, founded upon a theoretical analysis and regression analysis of experimental data [2,3] for structural design in seismic engineering. Priestley et al. [4] examined existing design equations related to shear strength, and observed significant differences in the predicted results. Sezen and Moehle [5] and Elwood and Moehle [6] also found significant inaccuracies in the predicted deformation capacity of RC columns obtained from existing methods. The reason for the differences is that RC material exhibits strong uncertainties and nonlinearity. Moreover, under the combination of constant vertical and lateral dynamic loads, the seismic performance of RC columns is affected by many other properties of structural components, such as the shear span ratio [7], longitudinal and transverse reinforcement ratio [8,9] and the axial compression ratio [10].

Besides, finite element analysis (FEA) is an alternative and useful tool that can be used to analyze the nonlinear mechanical properties of the RC structure [11,12]. For RC columns, the lumped plasticity (LP) hinge method and distributed plasticity (DP) hinge method are typical methods to model the nonlinear characteristics of the structures [13,14]. However, the LP method is relatively coarse, while the DP method is often time-consuming, and the results significantly depend upon the setup of the model parameters and boundary conditions [15]. This may lead to large estimation errors during the seismic performance prediction of RC structures. Lu et al. [16] organized an open competition to predict the hysteretic response of a 3-storey frame, which was initially tested in the laboratory. Although detailed information of the structural configuration and material properties was provided to the competitors, most of the FEA results of 30 research groups were far from the test results, no matter what types of FEA model they used. This may result from the complicated damage mechanism of the material and the unclear boundaries and connections in the real RC structures. To overcome these drawbacks, Ibarra et al. [17] developed a relatively simple LP model including the effective stiffness, pre and post-peak inelastic deformations, etc. It has been demonstrated that this model can simulate the seismic performance of RC structures with proper model parameters [18,19]. In practice, the accurate analysis is not only dependent upon the tools used, but also the experience of the analysts (mainly from the disaster lessons or the laboratory test results). The experience, which is generally related to the model parameter determination, may vary greatly among different groups. Therefore, essential model parameters considering the historical experience are needed for the accurate simulation of RC columns. In this study, a model-free and data-driven method is deduced for the determination of the key parameters of the commonly-used LP model. It is expected that more test results can overcome subjective errors and model uncertainties, improving the LP model.

An artificial neural network (ANN) is a typical machine learning (ML) method, which was inspired by the architecture and operation of biological nervous systems. Recently, they have been adopted by researchers in civil engineering, such as investigating energy performances [20]. Kawashima and Oreta [21] validated the application of ANN models in simulating the compressive stress–strain relationship, especially with limited data. In addition, other non-parametric models, the conditional average estimator (CAE) method [22] and support vector machine (SVM) [23], were also employed by researchers to analyze the mechanical properties or seismic performance of structures. González and Zapico [24] presented a seismic damage identification method for a steel moment-frame structure based on modal variables, and Reza et al. [25] conducted a detailed parametric study on the limiting states of bridge columns using factorial analysis. Compared with current existing formulas based on the regression algorithm or experiences, the ANN-based model, originating from an advanced humanlike information processing style, is data-driven. It only relies on a large quantity of training data with sufficient features of the samples rather than a limited database and certain assumptions, and thus it can help to reduce some subjective and experimental errors and be more reliable. Nevertheless, their application in engineering is still limited, and so this study conducted a pioneer study on developing an ANN-based LP model for the seismic assessment of RC columns. It investigates the possibility of predicting the important model properties of RC columns using an ANN model and a large quantity of historical test results.

Data is fundamental to the application of machine learning technology. The accumulation of numerous cyclic experiments in RC columns can provide an alternative approach to predict their seismic performance using a model-free method. In this study, the data collection criteria and experimental data for training, testing and validating the ANN model of the RC columns are briefly introduced in the first Section. Then, the ANN architecture, including the input, output and hidden layer, is described specifically. Subsequently, the ANN model is validated using a test set and some existing explicit formulas developed by other researchers. Finally, a comparison investigation is conducted on two RC frame structures between the ANN-based LP method with a quasi-static and a shaking table test. This well-trained, ANN-based LP method in this paper is also implemented in an efficient Matlab graphical user interface (GUI), which could be directly used for structural performance evaluation or response prediction by other researchers or engineers during their investigation.

2. Data Collection

Researchers worldwide have conducted numerous pseudo-static cyclic tests on RC columns to investigate the hysteretic behavior and seismic performance of such structural components in buildings and bridges. To provide data for training the ANN model, the cyclic test results of RC columns were collected from the NEEShub database [26], PEER database [27] and other published studies based upon the following criteria:

1. The shape of the cross section of the RC columns should be rectangular; others like circular columns are not included in this study.
2. The RC columns only sustained constant axial loading and unidirectional cyclic lateral force. Cyclic tests of the RC column under biaxial lateral or variable axial loads are excluded.
3. A complete loading process was applied to the specimens until failure; the load carrying capacity of the specimen decreased by more than 20% compared to the peak strength.
4. Details of the specimens are available, such as the geometrical size and reinforcement, as well as experimental results of the hysteretic curves.
5. Normal concrete was used in the manufacture of the specimens, without additives such as steel fiber.

A total of 1163 cyclic test results of RC columns were collected to build a database for training, testing and validating the ANN model. Details of the selected specimens, including the data source and geometrical size, are listed in the Appendix A. It can be observed that the main features of the selected columns have a relatively wide distribution.

The numerical simulation of RC columns may depend upon many factors, such as the material properties, geometrical configuration, reinforcement layout and loading protocol. Figure 1 shows a schematic diagram of a typical cantilever column, illustrating some of these factors, as well as its LP model. The section width (B) and depth (D) is defined as the section size which is perpendicular and parallel to the lateral force, respectively. It should be noted that a cantilever and fixed-end columns are two typical structural components used for civil structures. For simplicity, this study mainly focused on cantilever columns, which are widely used in bridge piers; the geometrical size and test results of the fixed-end columns were modified by assuming that the inflection point of the columns occurs at the half height of the column. As illustrated, the optimal feature subset is selected by the genetic algorithm (GA) method, then used to train the ANN model. It can help determine the parameters of the LP model for RC columns.

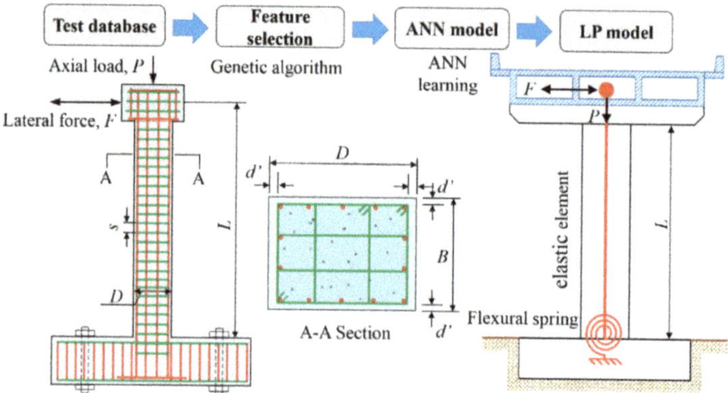

Figure 1. Schematic diagram and lumped plasticity model of a typical RC column.

As is known, there are many factors influencing the performance of RC columns, some of which are dependent. For example, the effective depth (d_e) can be calculated by B, D and the effective cover thickness (d'). Although the independent parameters are simple and easily accessible, the derived parameters may be more effective in some cases. Consequently, the potential influencing features are collected for feature selection, which are categorized into six groups as listed in Table 1. It can be observed from the table that a total of 24 features were included, in order to investigate their effects on RC columns.

Table 1. Features of reinforced concrete (RC) columns.

Category		Feature	Category		Feature
Cross Section	B	Width of column section	Long bars	ρ_l	Longitudinal reinforcement ratio
	D	Depth of column section		f_{yl}	Yield strength of longitudinal bar
	A_c	Gross area of column section		f_{ul}	Ultimate strength of longitudinal bar
	I_c	Moment of inertia of column section		d_l	Diameter of longitudinal bar
	d_e	Effective depth of column section		A_{sl}	Area of a single longitudinal bar
	d'	Effective cover thickness		N_l	Number of longitudinal bars
Span	L	Effective height of specimen		f_{yt}	Yield strength of transverse steel
	λ	Span-to-depth ratio		f_{ut}	Ultimate strength of transverse steel
Vertical loading	P	Applied axial load	Trans. bars	d_{st}	Diameter of transverse reinforcement bar
	n	Axial compression ratio		A_{st}	Area of one transverse reinforcement bar
Concrete	f_c'	28-day concrete compressive strength		s	Spacing of transverse reinforcement
	E_c	Elastic modulus of concrete		ρ_{sv}	Transverse reinforcement volumetric ratio

As for the modeling, the LP model of RC columns is usually represented by the bi-linear and tri-linear model [28]. As shown in Figure 2, the strength and the yielding, capping and ultimate deformation of the RC columns can be defined according to the method using the hysteretic curves obtained from cyclic tests. The strength is defined as the peak point when the resistance force reaches its maximum value. According to statistical results by Haselton et al. [29], the yielding deformation can be roughly estimated as that corresponding to 85% of the peak strength, and the ultimate deformation is obtained from the drift when the resistance drops to 20% of the peak strength [6]. It should be noted that the actual test results of the hysteretic curve cannot always achieve perfect symmetry in both the positive and negative directions. Therefore, the mean values of the strength and deformation in the positive and negative directions were adopted for the following analysis. The strength and deformation capacities of 1163 RC columns were determined according to the above approach, and used as the outputs of the ANN model.

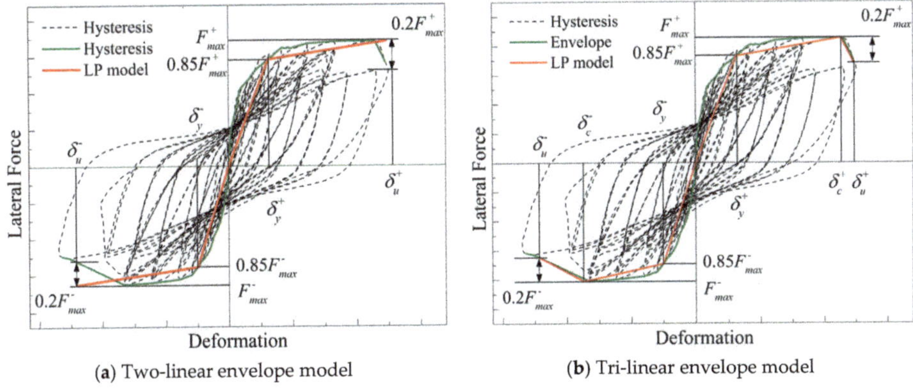

(**a**) Two-linear envelope model (**b**) Tri-linear envelope model

Figure 2. Schematic of hysteretic curves and definition of strength and deformation capacities.

Consequently, the model parameters of the LP model (e.g., yielding moment M_y, initial stiffness K_1, and the coefficient of the post yielding stiffness r) can be determined by the four parameters: The strength (F_{max}), the yielding (δ_y), capping (δ_c) and ultimate drift (δ_u), as shown in Tables 2 and 3. As can be seen, $M_y = F_yL + P\delta_yL$ and $M_u = F_{max}L + P\delta_cL$ are the yielding moment and ultimate moment, respectively. Consequently, the four variables (F_{max}, δ_u, δ_c, δ_y) are used as the outputs of the ANN model.

Table 2. Definition and determination of the bi-linear lumped plasticity (LP) model.

Parameter	Description	Calculation
K_e	the initial stiffness of the linear segment	M_y/δ_y
M_y	the yielding moment	$F_yL + P\delta_yL$
b	the hardening ratio	K_{sh}/K_e, $K_{sh} = (M_c - M_y)/(\delta_u - \delta_y)$

Table 3. Definition and determination of the tri-linear LP model [30].

Parameter	Description	Calculation
K_e	the initial stiffness of the linear segment	M_y/δ_y
M_y	the yielding moment	$F_yL + P\delta_yL$
β_l	the pre-capping hardening ratio	K_{sh}/K_e, $K_{sh} = (M_c - M_y)/(\delta_c - \delta_y)$
β_c	the post-capping hardening ratio	K_{ss}/K_e, $K_{ss} = (M_c - M_u)/(\delta_u - \delta_c)$
M_c/M_y	the capping moment to the yielding moment	$M_c = F_{max}L + P\delta_cL$, $M_u = F_{max}L + P\delta_cL$

3. ANN Model

3.1. Architecture of the ANN Model

ANN is a humanlike information processing method, which is composed of many highly interconnected processing elements or neurons. In this study, the developed ANN model that maps the features of the RC column and the structural capacities can be written as:

$$\mathbf{y} = f(\mathbf{x}; \boldsymbol{\theta}) \tag{1}$$

where $\mathbf{y}$ is the output feature vector of the ANN, including the strength and the yielding, capping and ultimate drift of the columns; $\mathbf{x}$ is the input feature vector of the column, as listed in Table 1; and $\boldsymbol{\theta}$ represents the learning parameters.

As shown in Figure 3, a typical ANN architecture consists of an input layer, output layer and one or several hidden layers, and each layer has corresponding neurons [31]. In this study, back propagation neural network (BPNN) [32] was adopted to predict the strength and deformation capacity of the columns. In the training process, the inputs propagate toward the output layer through the hidden layers, and errors between the predicted and experimental values back propagate from the output layer to the input layer to adjust the weights and thresholds of the hidden layers. In addition, the activation function defines how the input of a unit combines with its current activation level to complete a new activation. In this study, the commonly-used sigmoid activation function is utilized. Once the optimal connection weights and thresholds are determined, the trained ANN model can be conveniently employed to evaluate the strength and deformation capacities of RC columns.

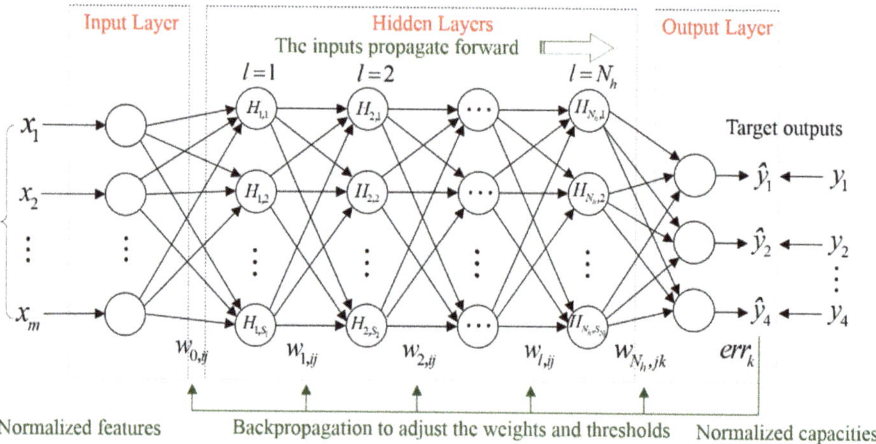

Figure 3. Architecture of the back propagation neural network (BPNN).

3.2. Input and Output Layer

As described previously, the strength and the deformation capacities of RC columns are affected by numerous factors. If all of the features are included in the input layer, the ANN model becomes very complex, and prone to fail due to overfitting. Therefore, it is reasonable to select an optimal input feature subset for the ANN model. To achieve optimization, all the input features listed in Table 1, and the output features are first organized into an $n \times m$ dimensional matrix and an $n \times 4$ dimensional matrix, respectively, as follows:

$$\mathbf{x} = \left\{x_{ij}\right\}_{n \times m}, \ \mathbf{y} = \left\{y_{ik}\right\}_{n \times 4} \tag{2}$$

where x_{ij} ($i = 1, 2, \cdots, n$) is the jth input feature of the ith RC column in the training set, and y_{ik} ($k = 1, 2, 3, 4$) is the corresponding output feature of the ith RC column. Considering the significant difference in the value ranges of the features, data obtained from the specimens and the test results of the training set are first normalized within the range of [0, 1] by:

$$\overline{x}_{ij} = \frac{x_{ij} - \min(\mathbf{x}_j)}{\max(\mathbf{x}_j) - \min(\mathbf{x}_j)}, \ \overline{y}_{ik} = \frac{y_{ik} - \min(\mathbf{y}_k)}{\max(\mathbf{y}_k) - \min(\mathbf{y}_k)} \tag{3}$$

where $\overline{x}_{ij}$ is jth normalized input feature corresponding to the ith specimen; $\mathbf{x}_j$ represents the jth input feature vector of all the specimens; and $\mathbf{y}_k$ denotes the kth output feature vector. After normalization, n input–output pairs ($\left\{\overline{\mathbf{x}}_i, \overline{\mathbf{y}}_i\right\}$, $i = 1, 2, \ldots, n$) are collected into the training set, where $\overline{\mathbf{x}}_i = \{\overline{x}_{i1}, \overline{x}_{i2}, \ldots, \overline{x}_{im}\}$ and $\overline{\mathbf{y}}_i = \{\overline{y}_{i1}, \overline{y}_{i2}, \ldots, \overline{y}_{i4}\}$ are the normalized input and output features of the ith column, respectively.

In this study, a genetic algorithm (GA) was adopted to find the optimal input features of the column. GA is a metaheuristic inspired by the process of theoretical Darwinian natural selection in biological systems to generate high-quality solutions of search problems [33]. In this method, each feature subset, such as structural configuration and material properties, is represented by an individual population. In the selection process, individuals with better phenotypic characteristics have greater probability to survive and reproduce in a population, whereas the less adapted individuals are more likely to disappear. Thus, the GA obtains the optimal solution after a series of iterative computations.

Figure 4 shows the chromosome design, fitness function and system architecture for the GA-based optimal feature selection of the columns, which operates in binary spaces (called chromosomes) and manipulates a population of potential solutions for the optimal input subset. A chromosome (genotype of the input features) is represented by binary coding as follows:

$$\boxed{g_1, g_2, \dots, g_j, \dots, g_m} \ , \quad g_j = \begin{cases} 1, & \text{if the } j\text{-th input feature is selected} \\ 0, & \text{if the } j\text{-th input feature is not selected} \end{cases} \tag{4}$$

where $g_1, g_2, \dots, g_m$ represent the 24 input features that will be selected. The initial chromosome population is randomly generated first, using binary coding. For example, chromosome 100100 means that the first and the fourth input features were selected.

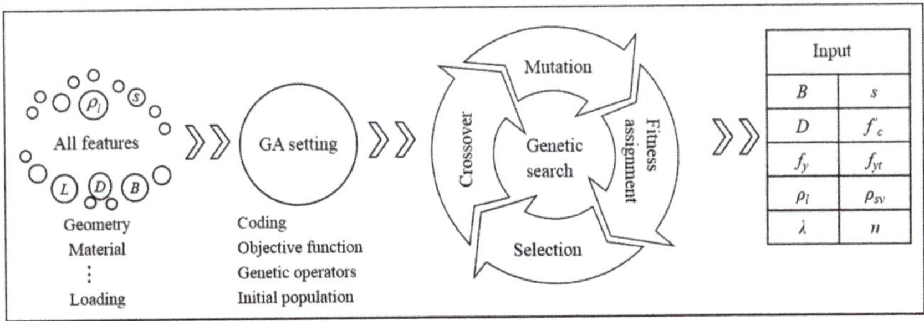

Figure 4. Genetic algorithm for optimal feature selection.

During the selection process, the fitness function is used to evaluate the quality of possible input subsets. There are several measurements can be used to evaluate the accuracy and survival probability of the chromosomes [34]. For convenience, the general fitness function is adopted in this study for the optimal input sets:

$$\text{fitness} = \frac{1}{\sum\limits_{k=1}^{4} \sum\limits_{i=1}^{n} \left(\overline{y}_{ik} - \hat{\overline{y}}_{ik}\right)^2} \tag{5}$$

where $\overline{y}_{ik}$ and $\hat{\overline{y}}_{ik}$ are the kth experimental result and predicted value of the ith specimen, respectively. For the selection of the optimal input features, an ANN model with one hidden layer containing 20 neurons is used, as the optimal ANN architecture is still not obtained in this stage.

Fitter chromosomes from the current population have a higher probability to be selected to generate offspring population using genetic operators, namely crossover and mutation. The population size, number of generations, crossover rate and mutation rate of the GA were selected as 500, 25, 0.8, and 0.3, respectively, using the approach proposed by Oliveira et al. [35]. This evaluation process will be performed iteratively until the termination criterion (10^{-3}) is satisfied. After that, the most fitted features can be determined as the inputs for ANN training.

The optimal input feature subset with 10 nodes, including B, D, f'_c, f_{yl}, λ, f_{yt}, s, ρ_l, ρ_{sv} and n were selected from the 24 features listed in Table 1 using GA. Laboratory test results reported by [36] also demonstrate that the selected optimal features are the most important parameters that affect the seismic performance of these RC columns. However, other parameters, like the effective cover thickness (d') and depth (d_e), may be also important. The reason for omitting them is that they are either similar in most specimens, or can be easily calculated. For example, the d' of most of the specimens is in the range of 0.02 m to 0.03 m, while the d_e can be determined based on B and d'.

4. Results and Discussion

4.1. Training of the ANN Model

Before training the ANN model, all of the collected structural and material parameters of the 1163 specimens and the corresponding test results on the strength and drift capacities were normalized as the input and output vectors using the above approach. Then, 863 specimens in the database were randomly selected as the training set ($n_{train} = 863$). While the remaining specimens were used as the test set ($n_{test} = 150$) and validation set ($n_{validation} = 150$), respectively.

The number of input and output nodes in the ANN architecture is obviously 10 and 4, respectively. The following parameters of the BPNN were used in training the model: (1) Error tolerance = 10^{-3}; (2) learning parameter = 0.15; (3) maximum number of iterations = 10^4; (4) momentum parameter = 0.05; (5) noise = 0.01. The mean absolute error (E) and goodness of fit (R^2) were adopted to evaluate the accuracy of the ANN model, as follows:

$$E_k = \frac{1}{N} \sum_{i=1}^{N} \left| \frac{\overline{y}_{ik} - \hat{\overline{y}}_{ik}}{\overline{y}_{ik}} \right|, \ (k = 1, 2, \ldots, 4) \tag{6}$$

$$R_k^2 = 1 - \frac{\sum_{i=1}^{N} \left(\overline{y}_{ik} - \hat{\overline{y}}_{ik}\right)^2}{\sum_{i=1}^{N} \left(\overline{y}_{ik}\right)^2}, \ (k = 1, 2, \ldots, 4) \tag{7}$$

where N is the number of specimens in the training set or test and validation set.

It should be noted that there is no specific rule or heuristics [37] to determine the number of hidden layers and the corresponding number of nodes. In general, an ANN model with too many neurons tends to fail due to an overfitting of the training data, whereas those with few neurons may not be able to capture the complex underlying relationship. Therefore, several trials were conducted for ANN models with different nodes and a single hidden layer to obtain an optimal ANN architecture for the strength and deformation capacity prediction of RC columns. For an optimal ANN architecture, the mean absolute error and goodness of fit of the ANN model should satisfy the following criteria: (1) Well-distributed around 0 (smaller E_k) and (2) as close as possible (larger R_k^2). The mean value ($E = \overline{E}_k$, $R^2 = \overline{R}_k^2$) of the four properties was used.

Figure 5 shows the test results of the trained ANN models with 10–50 hidden neurons. It can be observed that the mean absolute error and goodness of fit of the three properties are within the range of [0.126, 0.142] and [0.870, 0.882], respectively. As can be seen, the N-10-21-4 ANN architecture has the smallest E and largest R^2, and thus was determined as the optimal ANN architecture.

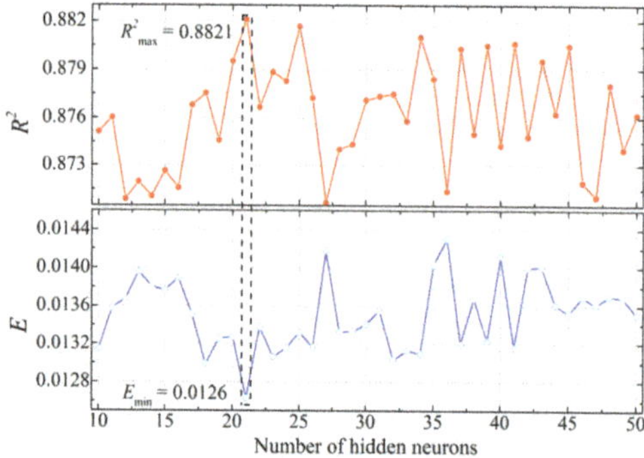

Figure 5. Comparison of artificial neural network (ANN) architectures.

4.2. Validation

Experimental data in the training set, test set and validation set were used to validate the ANN model. Figure 6 shows a comparison of the predicted strength capacities by the ANN model and the experiments. It can be observed that the predicted values of the strength are in good agreement with the experimental values of the columns. For the normalized strength, the mean absolute error and goodness of fit are 0.0533 and 0.9304, respectively, for the training set, and 0.0133 and 0.9431, respectively, for the test and validation set. The results indicate that the ANN model can be effectively used to predict the strength of the RC columns under seismic excitations.

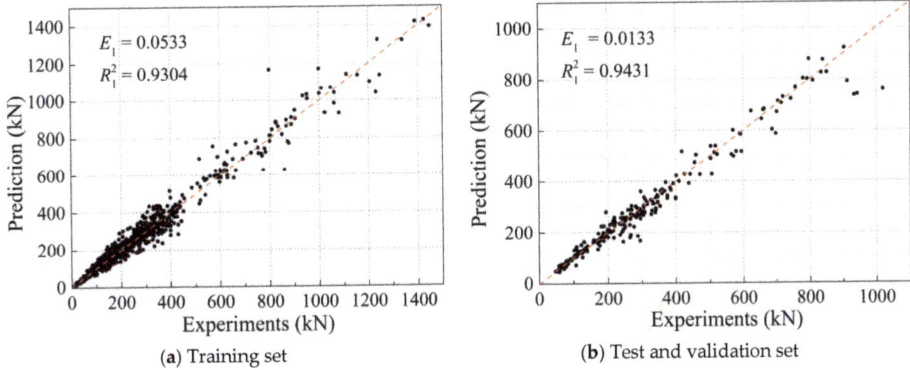

Figure 6. Comparison of experimental results and predicted results of the strength of RC columns.

Different types of explicit formulas have been proposed by researchers or adopted in design codes for calculating the strength of RC columns. In general, concrete and reinforcement contribute to the shear capacity of RC columns with shear failure mode. Therefore, the explicit formula has two distinct parts.

$$V = V_c + V_s \tag{8}$$

where $V = F_{\max}$ is the shear strength for shear-failed columns, while V_c and V_s are the contributions of concrete and steel reinforcement, respectively. In this study, three types of explicit formulas for predicting the shear strength of RC columns were adopted to validate the ANN model as summarized in Table 4.

Table 4. Explicit formulas for estimating the shear capacity of RC columns.

Reference	Concrete Contribution (V_c)	Steel Contribution (V_s)
Sezen and Moehle [5] FEMA 356 [3]	$V_c = k_\mu \left(\dfrac{0.5\sqrt{f_c'}}{\lambda} \sqrt{1 + \dfrac{P}{0.5\sqrt{f_c'}A_c}} \right) 0.8A_c$	$V_s = k_\mu \dfrac{A_{st}f_{yt}D\prime}{s}$
ACI 318-05 [38]	$V_c = \frac{1}{6}\left(1 + \dfrac{P}{13.8A_c}\right)\sqrt{f_c'}A_c$	$V_s = \dfrac{A_h f_{yh}D\prime}{s}$
FEMA 273 [39]	$V_c = 0.29\lambda\left(k + \dfrac{P}{13.8A_c}\right)\sqrt{f_c'}A_c$	$V_s = \dfrac{A_{st}f_{yt}D}{s}$

Note: k_μ is the ductility-related strength degradation factor, and D' is the distance from the extreme compression fiber to the centroid of the tension reinforcement. The meanings of other parameters in this table are the same as those in Table 1.

For flexural-dominated specimens, the lateral loading capacity is strongly dependent on the flexural strength (M_c) and the distance between the critical section of the plastic hinge and point of contra-flexure length (L), as follows:

$$F_{\max} = (M_c - P{\cdot}\Delta_c)/L = \left(M_y{\cdot}M_c/M_y - P{\cdot}\Delta_c\right)/L \tag{9}$$

where Δ_c is the capping displacement corresponding to the maximum lateral strength ($F_{\max}$). Fardis et al. [40] proposed a semi-empirical strength and deformation expression with good average agreement with test results. It has been widely used in the seismic performance assessment in engineering [29]

$$M_c / M_y = (1.25)(0.89)^n (0.91)^{0.01 f_c'} \tag{10}$$

$$M_y = BD^3 \phi_y \left\{ E_c \frac{k_y^2}{2} \left[0.5(1 + \delta') - \frac{k_y}{3} \right] + \frac{E_s}{2} \left[\left(1 - k_y\right)\rho + \left(k_y - \delta'\right)\rho' + \frac{\rho_v}{6}(1 - \delta') \right] (1 - \delta') \right\} \tag{11}$$

where ϕ_y is the yield curvature; δ' is the ratio of the distance of the compression reinforcement center from the extreme compression fibers to the span depth (D); k_y is the normalized compression zone depth; E_s is the reinforcement elastic modulus; ρ, ρ' and ρ_v are the reinforcement ratios of tension, compression and web reinforcement, respectively.

Figure 7 shows a comparison of the experimental results, predicted strength of the ANN model and calculated results from the explicit formula. In Figure 7a, the test results of the shear-failed specimens were obtained using the formula by [5], while the flexural failure specimens were selected from the database of this study. As shown in Figure 7a, the mean absolute error of the predicted shear strength of the ANN model is 0.107, which is smaller than those obtained from the three explicit formulas, 0.235, 0.207 and 0.217. In Figure 7b, it can also be observed that the mean absolute error of the ANN model is 0.087. However, the statistical result of the mean absolute error obtained from the design code is 0.101. It is illustrated that the results indicate that the ANN model yields a more accurate prediction of the structural strength of the columns.

Figure 8 also shows a comparison of the test results of the ultimate drift of the column, and predicted results from the ANN model. It can be observed that the ANN model can also achieve a reasonable prediction of the ultimate drift. However, it is evident that the mean absolute error of the ultimate drift is larger than that of the strength prediction, particularly for specimens with large ultimate deformation capacity.

The mean absolute error and goodness of fit are found to be 0.1761 and 0.8114, respectively, for the training set, and 0.1783 and 0.8216 for the test and validation set, respectively. This is because the ultimate deformation of the column is more strongly affected by the nonlinearity of the structures. Another reason for there being less accuracy is that the ultimate deformation is more difficult to measure.

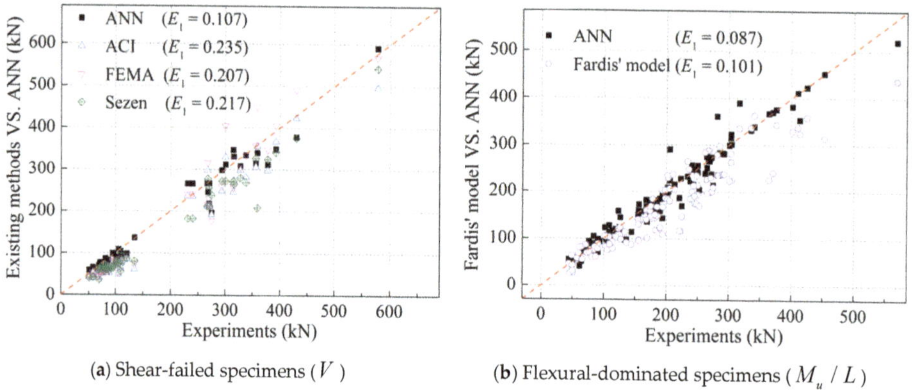

(**a**) Shear-failed specimens (V)   (**b**) Flexural-dominated specimens (M_u / L)

Figure 7. Comparison of the predicted results of the strength from the ANN model and existing explicit formulas.

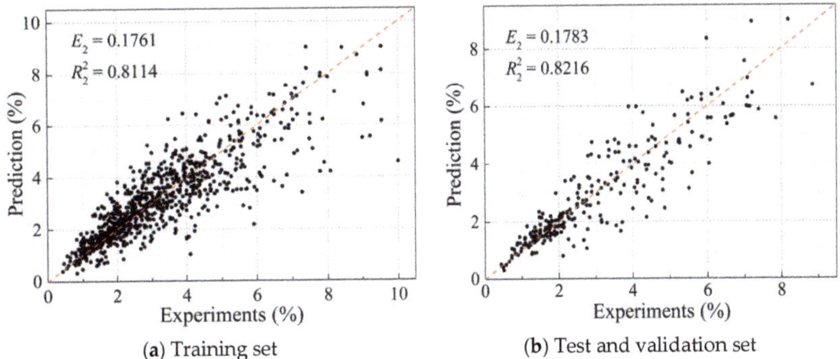

(a) Training set   **(b)** Test and validation set

Figure 8. Comparison of experimental results of the ultimate drift and predicted results from the ANN model.

To assess the ultimate deformation of lightly reinforced columns, Elwood et al. [6] proposed an explicit formula given as follows:

$$\frac{\Delta_u}{L} = \frac{3}{100} + 4\rho_t - \frac{1}{40}\frac{F_{max}}{A_c\sqrt{f_c'}} - \frac{n}{40} \geq 1\%(\text{MPa}) \tag{12}$$

where the effects of the transverse reinforcement ratio (ρ_t), axial load ratio (n) and strength (F_{max}) are taken into account. Furthermore, Lehman et al. [41] also developed an explicit method to evaluate the ultimate drift of flexural-dominated columns as follows:

$$\frac{\Delta_u}{L} = \frac{\Delta_y}{L} + \frac{\Delta_{sp}}{L} + \frac{\Delta_f}{L} \tag{13}$$

where Δ_y, Δ_{sp} and Δ_f are the yield, slip and flexural displacement, respectively.

The above explicit formulas for calculating the ultimate drift were also used to validate the accuracy and effectiveness of the ANN model. Figure 9 shows a comparison of results obtained from the formula methods and test results from the collected test database, which are divided into two parts, namely the shear-failed specimens and the flexural-dominated specimens. It can be observed that the ANN model also yields better prediction of the ultimate drift than the two explicit formulas. The mean absolute errors obtained from the ANN model and the Elwood's method are 0.165 and 0.266, respectively, for the shear-failure columns, whereas those obtained from the ANN model and the Lehman's method are 0.188 and 0.316, respectively, for the flexural-failure columns.

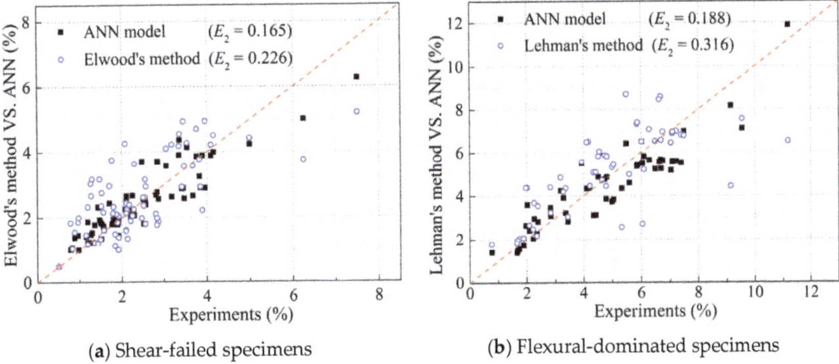

(a) Shear-failed specimens   **(b)** Flexural-dominated specimens

Figure 9. Comparison results of the ultimate drift between the experimental, ANN model and explicit methods.

Since there are currently no explicit formula for the estimation of capping drift corresponding to the maximum strength, Figure 10 only shows the comparison of the experimental results of the capping drift of the column and the predicted results from the ANN model, with the mean absolute error of 0.0794 and 0.1071, goodness of fit of 0.8001 and 0.8345, for the training set and the test and validation set, respectively. Nevertheless, it can still be seen that the ANN-based model for the capping drift estimation is reasonable and relatively accurate.

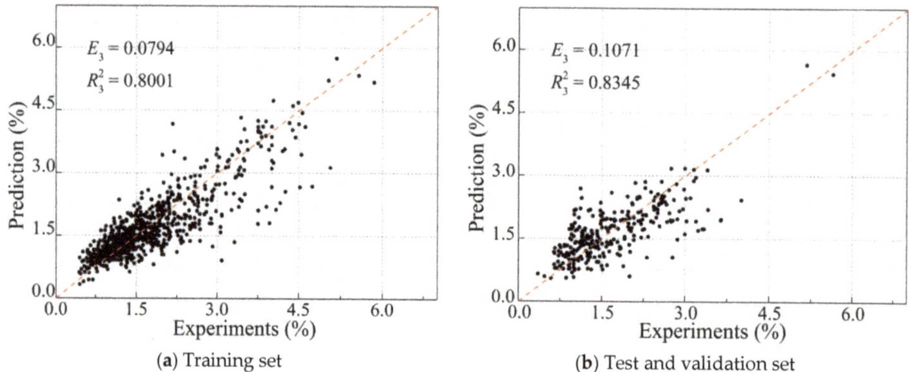

(**a**) Training set   (**b**) Test and validation set

Figure 10. Comparison of experimental results of the capping drift and predicted results from the ANN model.

Similarly, Figure 11 shows a comparison of the yielding drift of column between experimental results and predicted results from the ANN model. The mean absolute error for the training set and the test-validation set is 0.1081 and 0.0752, and the corresponding goodness of fit is 0.8548 and 0.8638, respectively. Generally, the yielding drift of RC columns always appears earlier with low nonlinearity than the ultimate drift. Therefore, the proposed ANN-based model can get a better prediction on the yielding drifts than the ultimate drift due to the nonlinearity difference between these two physical parameters.

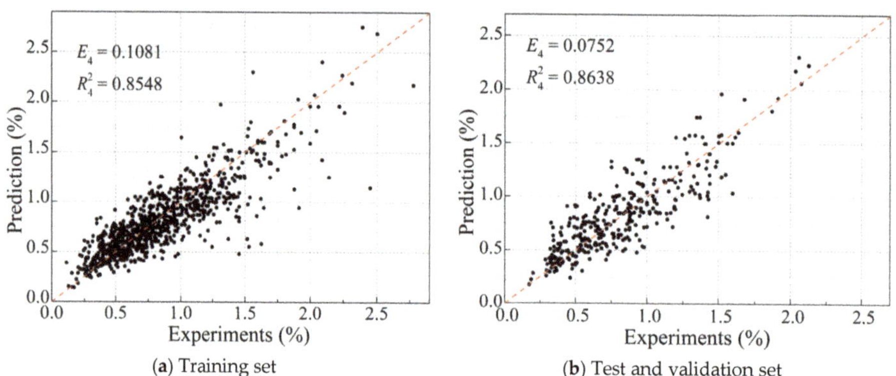

(**a**) Training set   (**b**) Test and validation set

Figure 11. Comparison of experimental results of the yielding drifts and predicted results from the ANN model.

For the estimation of the yielding drift, Fardis et al. [40] proposed a statistical formula based on the results of 963 tests, which has been widely used in engineering [29].

$$\delta_y = \phi_y \frac{L}{3} + 0.0025 + \alpha_{sl} \frac{0.25\varepsilon_y d_l f_{yl}}{(d - d')\sqrt{f_c'}} \tag{14}$$

where ϕ_y is the yielding curvature of the section, α_{sl} is the slip coefficient (equals 1 if slippage of longitudinal steel is possible, or 0 if it is not), and ε_y is the yield strain of longitudinal steel.

Figure 12 presents the comparison results of the yielding drift estimation of RC columns from the test set between the proposed ANN model and Fardis' method. It shows that the ANN model performs better than the empirical formula, with the mean absolute error of 0.2327 and 0.4572, respectively. Therefore, the developed ANN-based model can provide a more reliable and accurate prediction of the critical parameters for the LP model.

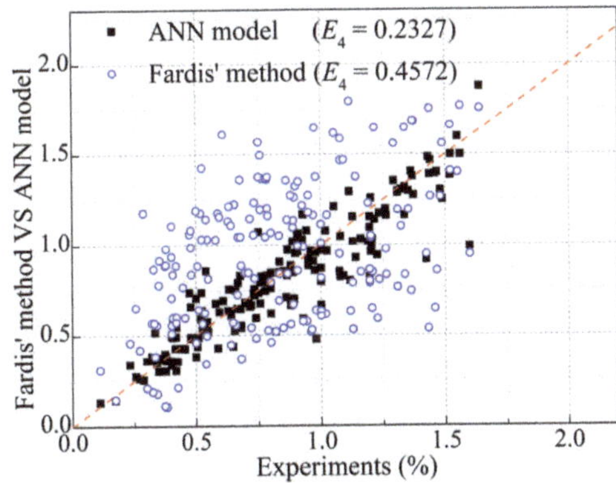

Figure 12. Comparison results of the yield drift between the experimental, ANN model and Fardis' method [40].

4.3. Evaluation of the ANN-Based LP Model

4.3.1. Comparison with the Pseudo-Static Test

The boundary conditions of the RC column during pseudo-static tests are generally close to its real situation. Therefore, it is reasonable to validate the ANN-based bi-linear and tri-linear LP model through a quasi-stable test on a frame. This test was conducted by [42], and the simulation is carried out using the Open System for Earthquake Engineering Simulation (OpenSEES, [28]) platform, as shown in Figure 13a. Each column was simulated by one elastic element and two zero-length elements at the two ends, as shown in Table 5. As illustrated in Figure 13b, the ANN-based LP model can provide an acceptable prediction result, especially for the envelope curve. It also shows that the ANN-based tri-linear model performs better than the bi-linear one during the whole hysteresis curve prediction process, especially after the capping point.

Table 5. ANN-based method for the LP model of the RC column sample [42].

Model	Parameters
ANN model	$F_{\max} = 27.50$ kN, $\delta_u = 4.53\%$, $\delta_c = 1.85\%$, $\delta_y = 1.17\%$
Bi-linear model	$K_e = 1119$ kN·m/rad, $M_y = 13.09$ kN·m, $b = 0.061$
Tri-linear model	$K_e = 1119$ kN·m/rad, $M_y = 13.09$ kN·m, $\beta_l = 0.18$, $\beta_c = -0.128$, $M_c/M_y = 1.17$

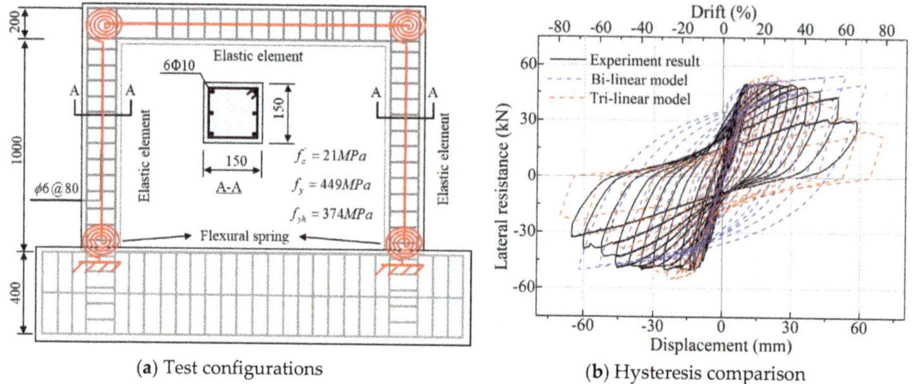

(**a**) Test configurations (**b**) Hysteresis comparison

Figure 13. Comparison of the ANN-based LP model and test result [42].

4.3.2. Comparison with Shake Table Test

For further validation, a time history response comparison was performed between the proposed ANN-based method and a shaking table test of a single-story RC frame [43]. As shown in Figure 14, the frame consists of two identical columns and a rigid link beam. Each column is modeled by one elastic element and two flexural springs at the two ends. Based on their properties, the ANN model was used to predict their strength and deformation capacities, as presented in Table 6. Therefore, the parameters for bilinear model (M_y, K_e, b) and tri-linear model parameters ($K_e, M_y, \beta_l, \beta_c, M_c/M_y$) of their LP models were calculated, based on which the FE model was developed on the OpenSEES platform.

During the test, the structure was subjected to a scaled ground motion with PGA = 1.28 g (recorded by TCU076NS in the 1999 Chi-Chi earthquake), as shown in Figure 15a. After performing the dynamic time history analysis, the drift response of the specimen was obtained and compared with the test result, as is shown in Figure 15b. It is observed that the ANN-based LP models can provide an acceptable estimation of the drift response before the collapse of the frame structure, especially for the tri-linear one. The corresponding maximum error of the drift response before the time of 22.6 s between the ANN-based tri-linear model and the test is very small. However, due to the limitation of the LP model itself, the collapse phenomenon of the specimen could not have been directly represented during the simulation. In spite of that, the proposed ANN-based model-free and data-driven method is still an effective and reliable candidate for the researchers and engineers to determine the key parameters of RC columns during finite element analysis.

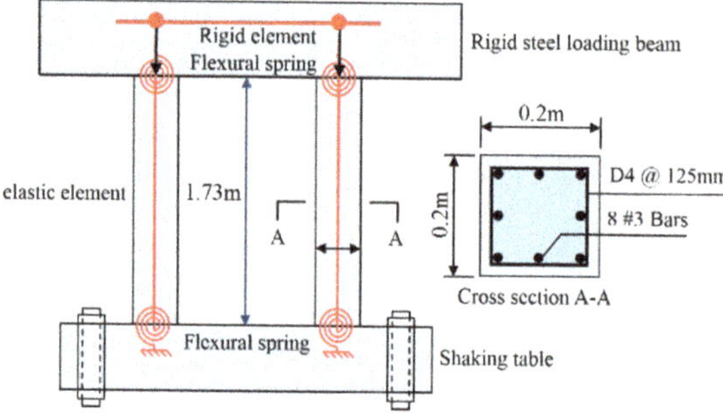

Figure 14. Prototype and model of the shaking table specimen.

Table 6. ANN-based method for the LP model of RC column sample S2 [43].

Model	Parameters
ANN model	$F_{max} = 36.62$ kN, $\delta_u = 4.53\%$, $\delta_c = 1.85\%$, $\delta_y = 0.99\%$
Bi-linear model	$K_e = 2720$ kN·m/rad, $M_y = 26.92$ kN·m, $b = 0.0494$
Tri-linear model	$K_e = 2720$ kN·m/rad, $M_y = 26.92$ kN·m, $\beta_l = 0.203$, $\beta_c = -0.087$, $M_c/M_y = 1.17$

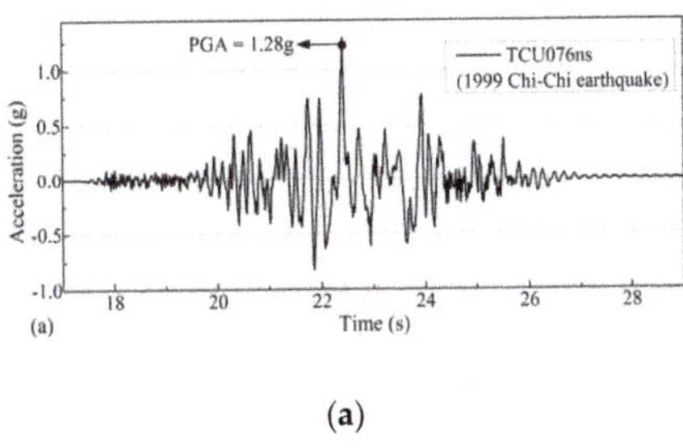

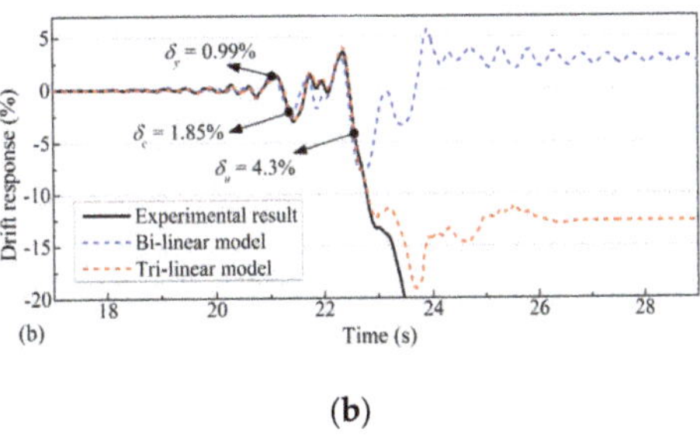

(a)   **(b)**

Figure 15. Drift response comparison of the frame obtained from test and ANN-based LP model: (a) Ground motion record; and (b) drift response.

The developed ANN-based model for the LP parameters of RC columns has been already implemented in an efficient Matlab GUI, anyone has access to it in the supplementary materials.

5. Conclusions

This study explored the possibility of using an ANN-based method to rapidly determine the seismic performance of RC columns, as well as the development of the LP model for them. An ANN model was established and validated based on the large database of 1163 cyclic tests of RC columns to predict the strength and the yielding, capping and ultimate deformation capacities, which are critical for the commonly-used LP model. The following conclusions are drawn:

1. On the basis of a large historical experimental database and advanced humanlike information processing algorithm, the proposed model-free method in this study can rapidly get more accurate input parameters for any bilinear and tri-linear LP model than current explicit formulas. The accuracy of the proposed method can also have been improved with the increment of the sample quantity of the database.
2. The validation results through both the collected experimental data and several existing functions indicate that the ANN-based method can be effectively used to predict the most important characteristics of RC columns, which are also critical for the further modeling of structures. In addition, another advantage of the proposed model-free method is that the quantity of the input features could be easily changed according to the requirement of an arbitrary multi-linear model.
3. The ANN-based LP model can help reduce the subjective and experimental errors. The prediction results of the RC frame structures using the well-trained ANN-based LP model show a good agreement with both the quasi-static and shaking table test results, especially for the pre-collapse stage. Thus, the model-free method based on the machine learning theory will be an innovative and promising approach for a fast seismic performance evaluation of the buildings and bridges in the future.

Thus, the model-free method based on the machine learning theory (e.g., ANN method) can be used as a promising surrogate for the rapid seismic performance evaluation of the buildings and bridges in the future.

Appl. Sci. **2019**, *9*, 4263

Supplementary Materials: The well-trained ANN-based model for the LP parameters of RC columns has been already implemented in an efficient Matlab GUI, which is available for the users, including both of the researchers and engineers to quickly evaluate the seismic performance of RC columns as well as RC structures. Anyone who is interested in the ANN-based LP model of RC column in this study may contact the author or download the program from the following address: https://www.dropbox.com/s/8aq3qreiochn26o/ANNLPGUI.zip?dl=0.

Author Contributions: Z.L. collected the test database, developed the ANN model and wrote the manuscript. S.L. contributed to the idea of the study, modified the computer program and supervised and verified the manuscript.

Funding: This research was funded by the National Key Research and Development Program of China (2018YFC0809404 and 2016YFC0701107) and the National Natural Science Foundation of China (51725801).

Conflicts of Interest: The authors declare no conflict of interest.

Appendix A

Table A1. Details of specimens included in the database.

References	Number of Specimens	Width (mm)	Shear Span (mm)	Concrete Strength (MPa)	Axial Load Ratio
Berry and Eberhard [44]	132	150~550	160~2200	16~160	0.0~0.8
Browning et al. [26]	168	80~800	80~2500	13~116	0.0~0.9
Yun [45]	6	510	1778	62.1~64.1	0.20~0.34
Ho and Johnny [46]	20	325	1515	56.5~111.1	0.11~0.55
Ongsupankul et al. [47]	4	400	1550	29.61~32.36	0.07~0.08
Woodward and Jirsa [48]	5	300	455	31~41	0.0~0.21
Bayrak [49]	24	250~350	1473	70.8~112.1	0.31~0.53
Mo and Wang [50]	9	400	1400	24.9~27.5	0.1~0.21
Paultre et al. [51]	8	305	2150	78.7~110	0.35~0.53
Xiao and Yun [52]	6	510	1778	62.1, 64.1	0.2~0.34
Lam et al. [53]	9	160~267	400~480	42, 47	0.4~0.65
Hwang and Yun [54]	8	200	300	68.3~70.3	0.3
Moretti and Tassios [7]	8	250	250~750	35~49	0.3,0.6
Ahn and Shin [55]	20	240	500	32~70	0.3~0.5
Woods et al. [56]	7	203	625	69	0.16
Marefat et al. [57]	7	150~200	750	20~28	0.16~0.31
Xiao et al. [58]	6	200	850	60, 90	0.38~0.54
Bae [59]	5	610, 440	2630	29.6~43.4	0.2, 0.5
Cao [60]	10	250, 350	600, 850	22.6~32.5	0.2~0.5
Ou et al. [61]	8	600	900	92.5~121	0.1, 0.2
Abdelsamie et al. [62]	7	250	700, 1050	26.6~151.4	0
Martirossyan and Xiao [63]	6	254	508	76, 86	0.1, 0.2
Li et al. [64]	8	300	250~500	23.4~27.5	0.09~0.29
Nakamura et al. [65]	6	450	450, 700	25, 28	0.16~0.18
Popa et al. [66]	7	300	450	18~29	0.2~0.4
Jin et al. [67]	8	150	495~660	34~73	0.09, 0.13
Bechtoula et al. [68]	10	325~520	813~1300	80, 130	0.3
El-Attar et al. [69]	7	150	870	141	0~0.35
Personal communications	626	150~900	150~3500	20~180	0~0.9

References

1. Guo, A.; Liu, Z.; Li, S.; Li, H. Seismic performance assessment of highway bridge networks considering post-disaster traffic demand of a transportation system in emergency conditions. *Struct. Infrastruct. Eng.* **2017**, *13*, 1523–1537. [CrossRef]
2. *Eurocode 8: Design of Structures for Earthquake Resistance—Part 3: Assessment and Retrofitting of Buildings;* European Committee for Standardization: Brussels, Belgium, 2005.
3. American Society of Civil Engineers. *Prestandard and Commentary for the Seismic Rehabilitation of Buildings;* FEMA 356: Washington, DC, USA, 2000.

4. Priestley, M.N.; Verma, R.; Xiao, Y. Seismic shear strength of reinforced concrete columns. *J. Struct. Eng.* **1994**, *120*, 2310–2329. [CrossRef]

5. Sezen, H.; Moehle, J.P. Shear Strength Model for Lightly Reinforced Concrete Columns. *J. Struct. Eng.* **2004**, *130*, 1692–1703. [CrossRef]

6. Elwood, K.J.; Moehle, J.P. Drift Capacity of Reinforced Concrete Columns with Light Transverse Reinforcement. *Earthq. Spectra* **2005**, *21*, 71–89. [CrossRef]

7. Moretti, M.; Tassios, T.P. Behaviour of short columns subjected to cyclic shear displacements: Experimental results. *Eng. Struct.* **2007**, *29*, 2018–2029. [CrossRef]

8. Taheri, A.; Moghadam, A.S.; Tasnimi, A.A. Critical factors in displacement ductility assessment of high-strength concrete columns. *Int. J. Adv. Struct. Eng.* **2017**, *9*, 325–340. [CrossRef]

9. Mostafaei, H.; Vecchio, F.J.; Kabeyasawa, T. Deformation Capacity of Reinforced Concrete Columns. *ACI Struct. J.* **2009**, *106*, 187–195.

10. Wibowo, A.; Wilson, J.L.; Lam, N.T.K.; Gad, E.F. Drift Capacity of Lightly Reinforced Concrete Columns. *Aust. J. Struct. Eng.* **2011**, *15*, 131–150. [CrossRef]

11. Ferdous, W.; Manalo, A.; Aravinthan, T.; Fam, A. Flexural and shear behaviour of layered sandwich beams. *Const. Build. Mater.* **2018**, *173*, 429–442. [CrossRef]

12. Ferdous, W.; Manalo, A.; Erp, G.V.; Aravinthan, T.; Ghabraie, K. Evaluation of an Innovative Composite Railway Sleeper for a Narrow-Gauge Track under Static Load. *J. Compos. Const.* **2018**, *22*, 04017050. [CrossRef]

13. Kim, S.-H.; Han, S.-J.; Kim, K.S. Nonlinear Finite Element Analysis Formulation for Shear in Reinforced Concrete Beams. *Appl. Sci.* **2019**, *9*, 3503. [CrossRef]

14. Szcześniak, A.; Stolarski, A. Dynamic Relaxation Method for Load Capacity Analysis of Reinforced Concrete Elements. *Appl. Sci.* **2018**, *8*, 396. [CrossRef]

15. Lucchini, A.; Franchin, P.; Kunnath, S. Failure simulation of shear-critical RC columns with non-ductile detailing under lateral load. *Earthq. Eng. Struct. Dyn.* **2017**, *46*, 855–874. [CrossRef]

16. Lu, X.; Ye, L.; Pan, P.; Zhao, Z.; Ji, X.; Qian, J. Pseudo-static collapse experiments and numerical prediction competition of RC frame structure I:RC frame experiment. *Build. Struct.* **2012**, *42*, 23–26.

17. Ibarra, L.F.; Medina, R.A.; Krawinkler, H. Hysteretic models that incorporate strength and stiffness deterioration. *Earthq. Eng. Struct. Dyn.* **2005**, *34*, 1489–1511. [CrossRef]

18. Haselton, C.B.; Liel, A.B.; Deierlein, G.G. Simulation structural collapse due to earthquakes: Model calibration, and numerical solution algorithms. In Proceedings of the Computational Methods in Structural Dynamics and Earthquake Engineering (COMPDYN), Rhodes, Greece, 22–24 June 2009.

19. Liel Abbie, B.; Haselton Curt, B.; Deierlein Gregory, G. Seismic Collapse Safety of Reinforced Concrete Buildings. II: Comparative Assessment of Nonductile and Ductile Moment Frames. *J. Struct. Eng.* **2011**, *137*, 492–502. [CrossRef]

20. Lucchi, E. Non-invasive method for investigating energy and environmental performances in existing buildings. In Proceedings of the PLEA—Architecture and Sustainable Development, Louvain-la-Neuve, Belgium, 13–15 July 2011.

21. Kawashima, K.; Oreta, A.W.C. Neural Network Modeling of Confined Compressive Strength and Strain of Circular Concrete Columns. *J. Struct. Eng.* **2003**, *129*, 554–561.

22. Iztok, P.; Karmen, P.; Fajfar, P. Flexural deformation capacity of rectangular RC columns determined by the CAE method. *Earthq. Eng. Struct. Dyn.* **2006**, *35*, 1453–1470.

23. Chou, J.-S.; Ngo, N.-T.; Pham, A.-D. Shear Strength Prediction in Reinforced Concrete Deep Beams Using Nature-Inspired Metaheuristic Support Vector Regression. *J. Comput. Civ. Eng.* **2016**, *30*, 04015002. [CrossRef]

24. González, M.P.; Zapico, J.L. Seismic damage identification in buildings using neural networks and modal data. *Comput. Struct.* **2008**, *86*, 416–426. [CrossRef]

25. Reza, S.M.; Alam, M.S.; Tesfamariam, S. Lateral load resistance of bridge piers under flexure and shear using factorial analysis. *Eng. Struct.* **2014**, *59*, 821–835. [CrossRef]

26. Browning, J.A.; Pujol, S.; Eigenmann, R.; Ramirez, J.A. NEEShub Databases. *Concrete International.* 2013, Volume 35. Available online: https://datacenterhub.org/resources/databases (accessed on 4 April 2013).

27. University of Washington. The UW-PEER Reinforced Concrete Column Test Database. Available online: http://www.ce.washington.edu/$\sim$peera1/: (accessed on 1 January 2004).

28. Mckenna, F.; Fenves, G.L. Open system for earthquake engineering simulation (OpenSees). In *Pacific Earthquake Engineering Research Center*; University of California: California, CA, USA, 2013.

29. Haselton, C.B. *Beam-Column Element Model Calibrated for Predicting Flexural Response Leading to Global Collapse of RC Frame Buildings*; Pacific Earthquake Engineering Research Center: Berkeley, CA, USA, 2008.

30. Dimitrios, L. *Sidesway Collapse of Deteriorating Structural Systems under Seismic Excitations*; Stanford University: Standford, CA, USA, 2009.

31. Ziegel, E.R. *The Elements of Statistical Learning*; World Publishing Corporation: New York, NY, USA, 2015.

32. Flood, I. *Neural networks in civil engineering: A review. Civil and Structural Engineering Computing*; Saxe-Coburg Publications: Stirling, UK, 2001; pp. 185–209.

33. Booker, L.B.; Goldberg, D.E.; Holland, J.H. Classifier systems and genetic algorithms. *Artif. Intell.* **1989**, *40*, 235–282. [CrossRef]

34. Shafti, L.S.; Pérez, E. Fitness Function Comparison for GA-Based Feature Construction. In Proceedings of the Current Topics in Artificial Intelligence, Conference of the Spanish Association for Artificial Intelligence, Caepia, Salamanca, Spain, 12–16 November 2007; pp. 249–258.

35. Oliveira, A.L.I.; Braga, P.L.; Lima, R.M.F.; Lio, M.; Rcio, L. GA-based method for feature selection and parameters optimization for machine learning regression applied to software effort estimation. *Inf. Softw. Technol.* **2010**, *52*, 1155–1166. [CrossRef]

36. Li, D.; Jin, L.; Du, X.; Fu, J.; Lu, A. Size effect tests of normal-strength and high-strength RC columns subjected to axial compressive loading. *Eng. Struct.* **2016**, *109*, 43–60. [CrossRef]

37. Zhang, G.; Patuwo, B.E.; Hu, M.Y. Forecasting with artificial neural networks: The state of the art. *Int. J. Forecast.* **1998**, *14*, 35–62. [CrossRef]

38. ACI 318-05. *Building Code Requirements for Structural Concrete*; Farmington Hills: Oakland, MI, USA, 2002; pp. 16–17.

39. FEMA, F. *NEHRP Guidelines for the Seismic Rehabilitation of Buildings: FEMA 273*; Federal Emergency Management Agency: Washington, DC, USA, 1997.

40. Telemachos, B.P.; Michael, N.F. Deformations of Reinforced Concrete Members at Yielding and Ultimate. *Aci Struct. J.* **2001**, *98*, 135–147. [CrossRef]

41. Moehle, J.P.; Lehman, D.E. Seismic Performance of Well-confined Confined Concrete Bridge Columns. *Spec. Publ.* **2000**, *238*. [CrossRef]

42. Tekeli, H.; Aydin, A. An experimental study on the seismic behavior of infilled RC frames with opening. *Sci. Iran.* **2017**. [CrossRef]

43. Wu, C.L.; Yang, Y.S.; Hwang, S.J.; Loh, C.H. Dynamic collapse of reinforced concrete columns. In Proceedings of the 9th U.S. National and 10th Canadian Conference on Earthquake Engineering, Toronto, ON, Canada, 25–29 July 2010.

44. Berry, M.; Eberhard, M. *Performance Models for Flexural Damage in Reinforced Concrete Columns*; University of Califorlia Berkeley: Berkeley, CA, USA, 2003.

45. Yun, H.W. Full-Scale Experimental and Analytical Studies on High-Strength Concrete Columns. Ph.D. Thesis, University of Southern California, Los Angeles, CA, USA, 2003.

46. Ho, J.C.M. Inelastic Design of Reinforced Concrete Beams and Limited Ductilehigh-Strength Concrete Columns. Ph.D. Thesis, The University of Hong Kong, Hong Kong, China, 2003.

47. Ongsupankul, S.; Kanchanalai, T.; Kawashima, K. Behavior of reinforced concrete bridge pier columns subjected to moderate seismic load. *ScienceAsia* **2007**, *33*, 175–185. [CrossRef]

48. Woodward, K.A.; Jirsa, J.O. Influence of Reinforcement on RC Short Column Lateral Resistance. *J. Struct. Eng.* **1984**, *110*, 90–104. [CrossRef]

49. Bayrak, O. Seismic Performance of Rectilinearly Confined High Strength Concrete Columns. Ph.D. Thesis, University of Toronto, Toronto, OT, Canada, 1999.

50. Mo, Y.L.; Wang, S.J. Seismic Behavior of RC Columns with Various Tie Configurations. *J. Struct. Eng.* **2000**, *126*, 1122–1130. [CrossRef]

51. Paultre, P.; Légeron, F.; Mongeau, D. Influence of concrete strength and transverse reinforcement yield strength on behavior of high-strength concrete columns. *ACI Struct. J.* **2001**, *98*, 490–501.

52. Xiao, Y.; Yun, H.W. Experimental studies on full-scale high-strength concrete columns. *ACI Struct. J.* **2002**, *99*, 199–207.

53. Lam, S.S.E.; Wang, Z.Y.; Liu, Z.Q.; Wong, Y.L.; Li, C.S.; Wu, B. Drift Capacity of Rectangular Reinforced Concrete Columns with Low Lateral Confinement and High-Axial Load. *J. Struct. Eng.* **2003**, *129*, 733–742. [CrossRef]

54. Hwang, S.K.; Yun, H.D. Effects of transverse reinforcement on flexural behaviour of high-strength concrete columns. *Eng. Struct.* **2004**, *26*, 1–12. [CrossRef]

55. Ahn, J.M.; Shin, S.W. An evaluation of ductility of high-strength reinforced concrete columns subjected to reversed cyclic loads under axial compression. *Mag. Concr. Res.* **2007**, *59*, 29–44. [CrossRef]

56. Woods, J.M.; Kiousis, P.D.; Ehsani, M.R.; Saadatmanesh, H.; Fritz, W. Bending ductility of rectangular high strength concrete columns. *Eng. Struct.* **2007**, *29*, 1783–1790. [CrossRef]

57. Marefat, M.S.; Khanmohammadi, M.; Bahrani, M.K.; Goli, A. Experimental Assessment of Reinforced Concrete Columns with Deficient Seismic Details under Cyclic Load. *Adv. Struct. Eng.* **2008**, *9*, 337–347. [CrossRef]

58. Xiao, X.; Guan, F.L.; Yan, S. Use of ultra-high-strength bars for seismic performance of rectangular high-strength concrete frame columns. *Mag. Concr. Res.* **2008**, *60*, 253–259. [CrossRef]

59. Bae, S. Seismic performance of full-scale reinforced concrete columns. *Aci Struct. J.* **2008**, *105*, 123–133.

60. Cao, T.N.T. Experimental and Analytical Studies on the Seismic Behavior of Reinforced Concrete Columns with Light Transverse Reinforcement. Ph.D. Thesis, Nanyang Technological University, Singapore, 2010.

61. Ou, Y.C.; Kurniawan, D.P.; Handika, N. Shear Behavior of Reinforced Concrete Columns with High-Strength Steel and Concrete under Low Axial Load. *ACI Mater. J.* **2015**, *112*, 35–45.

62. Abdelsamie, E.; Tom, B.; Salah, E. Plastic Hinge Length Considering Shear Reversal in Reinforced Concrete Elements. *J. Earthq. Eng.* **2012**, *16*, 188–210.

63. Martirossyan, A.; Xiao, Y. Flexural-Shear Behavior of High-Strength Concrete Short Columns. *Earthq. Spectra* **2012**, *17*, 679–695. [CrossRef]

64. Li, Y.A.; Huang, Y.T.; Hwang, S.J. Seismic Response of Reinforced Concrete Short Columns Failed in Shear. *Aci Struct. J.* **2014**, *111*, 945–954. [CrossRef]

65. Nakamura, T.; Yoshimura, M. Gravity load collapse of reinforced concrete columns with decreased axial load. In Proceedings of the 2nd European Conference on Earthquake Engineering and Seismology, Istanbul, Turkey, 25–29 August 2014.

66. Popa, V.; Cotofana, D.; Vacareanu, R. Effective stiffness and displacement capacity of short reinforced concrete columns with low concrete quality. *Bull. Earthq. Eng.* **2014**, *12*, 2705–2721. [CrossRef]

67. Jin, C.; Pan, Z.; Meng, S.; Qiao, Z. Seismic Behavior of Shear-Critical Reinforced High-Strength Concrete Columns. *J. Struct. Eng.* **2015**, *141*, 04014198. [CrossRef]

68. Bechtoula, H.; Kono, S.; Watanabe, F.; Mehani, Y.; Kibboua, A.; Naili, M. Performance of HSC columns under severe cyclic loading. *Bull. Earthq. Eng.* **2015**, *13*, 503–538. [CrossRef]

69. El-Attar, M.M.; El-Karmoty, H.Z.; El-Moneim, A.A. The behavior of ultra-high-strength reinforced concrete columns under axial and cyclic lateral loads. *HBRC J.* **2016**, *12*, 284–295. [CrossRef]

MDPI

St. Alban-Anlage 66

4052 Basel

Switzerland

Tel. +41 61 683 77 34

Fax +41 61 302 89 18

www.mdpi.com

Applied Sciences Editorial Office

E-mail: applsci@mdpi.com

www.mdpi.com/journal/applsci